Solutions and Tests For Exploring Creation With General Science

Manufactured in the United States of America
Fourth Printing 2003

Published By

Apologia Educational Ministries, Inc.
Anderson, IN

Printed by

The C.J. Krehbiel Company
Cincinnati, OH

Exploring Creation With General Science

Solutions and Tests

TABLE OF CONTENTS

Answers to the Tests

TEACHER'S NOTES
Exploring Creation With General Science
Dr. Jay L. Wile

Thank you for purchasing *Exploring Creation With General Science*. I designed this modular course specifically to meet the needs of the homeschooling parent. I am very sensitive to the fact that most homeschooling parents do not know the upper-level sciences very well, if at all. As a result, they consider it nearly impossible to teach to their children. This course has several features that make it ideal for such a parent.

1. The course is written in a conversational style. Unlike many authors, I do not get wrapped up in the desire to write formally. As a result, the text is easy to read and the student feels more like he or she is *learning*, not just reading.

2. The course is completely self-contained. Each module includes the text of the lesson, experiments to perform, questions to answer, and a test to take. The solutions to the questions are fully explained, and the test answers are provided. The experiments are written in a careful, step-by-step manner that tells the student not only what he or she should be doing, but also what he or she should be observing.

3. Most importantly, this course is Christ-centered. In every way possible, I try to make science glorify God. One of the most important things that you and your student should get out of this course is a deeper appreciation for the wonder of God's Creation!

I hope that you enjoy using this course as much as I enjoyed writing it!

Pedagogy of the Text

(1) There are two types of exercises that the student is expected to complete: "on your own" questions, and an end-of-the module study guide.

- The "on your own" questions should be answered as the student reads the text. The act of answering these questions will cement in the student's mind the concepts he or she is trying to learn. Answers to these problems are in the student text

- The study guide should be completed in its entirety after the student has finished each module. Answers to the study guide questions are in this book.

The student should be allowed to study the solutions to the "on your own" questions while he or she is working on them. When the student reaches the study guide, however, the solutions should be used only to check the student's completed work.

(2) In addition to the solutions to the study guides, there is a test for each module in this book, along with the answers to the test. **I strongly recommend that you administer**

each test once the student has completed the module and all associated exercises. The student should be allowed to have only pencil, paper, and a calculator. I understand that many homeschoolers do not like the idea of administering tests. However, if your student is planning to attend college, it is *absolutely* necessary that he or she become comfortable with taking tests!

(3) The best way to grade the tests is to assign one point for every answer that the student must supply. Thus, if a question has three parts and an answer must be supplied for each part, the question should be worth 3 points. The student's percentage correct, then, is simply the number of answers the student got right divided by the total number of answers times 100. The student's letter grade should be based on a 90/80/70/60 scale.

(4) All definitions presented in the text are centered. The words will appear in the study guide and their definitions need to be memorized.

(5) Words that appear in bold-face type in the text are important terms that the student should know.

(6) The study guide gives your student a good feel for what I require him or her to know for the test. Any information needed to answer the study guide questions is information that the student must know for the test.

Experiments

The experiments in this course are designed to be done as the student is reading the text. I recommend that your student keep a notebook of these experiments. This notebook serves two purposes. First, as the student writes about the experiment in the notebook, he or she will be forced to think through all of the concepts that were explored in the experiment. This will help the student cement them into his or her mind. Second, certain colleges might actually ask for some evidence that your student did, indeed, have a laboratory component to his or her science course. The notebook will not only provide such evidence but will also show the college administrator the quality of the science instruction that you provided to your student. I recommend that you perform your experiments in the following way:

- When your student gets to the experiment during the reading, have him or her read through the experiment in its entirety. This will allow the student to gain a quick understanding of what her or she is to do.

- Once the student has read the experiment, he or she should then start a new page in his or her laboratory notebook. The first page should be used to write down all of the data taken during the experiments and perform any exercises discussed in the experiment.

- When the student has finished the experiment, he or she should write a brief report in his or her notebook, right after the page where the data and exercises were written. The report should be a brief discussion of what was done and what was learned. The discussion should be written so that someone who has never read the book can read the discussion and figure out what basic procedure was followed and what was learned as a result of the experiment.

- **PLEASE OBSERVE COMMON SENSE SAFETY PRECAUTIONS. The experiments are no more dangerous than most normal, household activity. Remember, however, that the vast majority of accidents do happen in the home!**

Question/Answer Service

For all those who use my curriculum, I offer a question/answer service. If there is anything in the modules that you do not understand - from an esoteric concept to a solution for one of the problems - just get in touch with me by any of the means listed on the **NEED HELP?** page that is in the student textbook.

Solutions To The

Study Guides

SOLUTIONS TO THE STUDY GUIDE FOR MODULE #1

1. a. <u>Science</u> - A branch of study dedicated to the accumulation and classification of observable facts in order to formulate general laws about the natural world

b. <u>Papyrus</u> - A primitive form of paper, made from a long-leafed plant of the same name

c. <u>Spontaneous generation</u> - The idea that living organisms can be spontaneously formed from non-living substances

2. a. <u>We should support a scientific idea based on the evidence, not based on the people who agree with it</u>. Belief in spontaneous generation and the Ptolemaic system lasted so long mostly out of respect for Aristotle and Ptolemy, not because of the evidence.

b. <u>Scientific progress depends not only on scientists, but also on government and culture</u>. Science stalled in the Dark Ages because there was little government and cultural support for it.

c. <u>Scientific progress occurs by building on the work of previous scientists</u>.

3. <u>Imhotep was an ancient Egyptian doctor</u>. His medical practices were renowned throughout the known world.

4. <u>The ancient Egyptians never used their observations to explain the world around them</u>. Instead, they simply took a trial and error approach to finding cures for illness, etc. True science requires observation *and* explanation.

5. <u>They were ancient Greeks who were considered to be the first three scientists</u>.

6. They are best remembered for their idea that all matter is comprised of <u>atoms</u>.

7. <u>Aristotle</u>

8. <u>Aristotle</u>

9. <u>The Ptolemaic system placed the earth at the center of the universe and had both the planets and the stars traveling around the earth. The Copernican system placed the sun at the center and had the planets traveling around it. The Copernican system is more correct</u>.

10. <u>They wanted to turn lead into gold</u>. "Inexpensive substances" can be substituted for "lead" and "expensive substances" could be substituted for gold.

11. <u>They were not true scientists because their approach was strictly trial and error</u>.

12. <u>Science began to progress towards the end of the Dark Ages because the Christian worldview began to replace the Roman worldview</u>. Since the Christian worldview is a perfect fit

with science, the establishment of that worldview was essential for starting scientific progress again.

13. Grosseteste was the first modern scientist because he was first to use the scientific method.

14. The authors were Copernicus and Vesalius. The book by Copernicus was about the arrangement of the stars and planets in space, and the book by Vesalius was on the human body.

15. Galileo was forced to recant belief in the Copernican system by the Roman Catholic church. He would have otherwise been thrown out of the church.

16. Galileo claimed to invent the telescope, but he actually stole the idea.

17. Newton is the single greatest scientist of all time. He laid down the laws of motion, developed a universal law of gravity, and invented calculus.

18. The good part of the change was that science began to stop relying on the authority of previous, great scientists. The bad part of the change was that science began to move away from the authority of the Bible.

19. Lavoisier came up with the law of mass conservation.

20. Dalton is remembered for the first detailed atomic theory.

21. Darwin is best known for his book, *Origin of Species*. You could also say evolution.

22. The immutability of species refers to the mistaken idea that living creatures cannot change. Darwin showed that this is just not true.

23. Gregor Mendel is remembered for his work on how traits are passed on during reproduction. You could also say genetics.

24. James Clerk Maxwell is known as the founder of modern physics.

25. James Joule came up with The First Law of Thermodynamics.

26. Max Planck made the assumption that energy comes in small packets called "quanta."

27. Niels Bohr is remembered for his theory of the atom.

28. Einstein also developed the special theory of relativity and the general theory of relativity.

SOLUTIONS TO THE STUDY GUIDE FOR MODULE #2

1. a. <u>Counter-example</u> - An example that contradicts a scientific conclusion

b. <u>Hypothesis</u> - An educated guess that attempts to explain an observation or answer a question

c. <u>Theory</u> - A hypothesis that has been tested with a significant amount of data

d. <u>Scientific law</u> - A theory that has been tested by and is consistent with generations of data

2. <u>Science can never prove anything</u>.

3. <u>No, it does not</u>. You saw that in Experiments 2.1 and 2.2. After all, the idea that heavier things fall faster than lighter things makes sense. Nevertheless, it is wrong!

4. <u>The penny will hit the ground first</u>. Remember, the fact that all things fall at the same rate is only true when there is no air. Air resistance will slow the feather down more than the penny.

5. <u>Neither will hit the bottom of the tube first, because they will both fall at the same rate</u>. Since there is no air in the tube, objects will fall at the same rate, regardless of their weight.

6. To destroy a scientific law, you need <u>only one counter-example</u>. Remember, a scientific law is established simply because the theory has been confirmed by an enormous amount of experimentation. If a single experiment can be demonstrated to contradict the law, it is no longer a law!

7. <u>The observation was that the objects similar to the one he was studying had been seen before by other scientists at regular intervals in history</u>.

8. <u>His hypothesis was that the object he was studying was the same thing that the other scientists had seen before</u>.

9. <u>The experiment was to confirm the presence of the comet again in 1758</u>.

10. Since the appearance of the comet has been noted many times throughout history by many different scientists, <u>the existence of Halley's comet is now a scientific law</u>.

11. c. <u>Make observations</u>
a. <u>Form a hypothesis</u>
e. <u>Perform experiments to confirm the hypothesis</u>
d. <u>Hypothesis is now a theory</u>
f. <u>Perform many experiments over several years</u>
b. <u>Theory is now a law</u>

12. <u>You can either discard the hypothesis or modify it to become consistent with the experiment</u>.

13. <u>You can either discard the theory or modify it to become consistent with the experiment</u>.

14. The observation that led to Lowell's hypothesis was the <u>fact that there were faint lines on the surface of Mars</u>. The experiments used to confirm the hypothesis were <u>Lowell's detailed studies of Mars' surface</u>.

15. The discovery of high-temperature superconductors was startling because <u>a generally-accepted scientific law said that it was impossible to have high-temperature superconductors</u>.

16. a. <u>It cannot prove anything</u>.
    b. <u>It is not 100% reliable</u>
    c. <u>It must conform to the scientific method</u>

17. <u>Yes, it can</u>. As long as the scientific method is followed, science can be used to study *anything*!

18. <u>Yes, it can</u>. As long as the scientific method is followed, science can be used to study *anything*!

19. The observations were that <u>many people draw strength, hope, and encouragement from the Bible</u>.

20. I hypothesized that <u>the Bible is the Word of God</u>.

21. <u>I searched the Bible for knowledge of future events</u>. This would indicate that the Creator of time itself had inspired the Book.

22. <u>Of course not</u>! Science cannot prove anything. I did confirm the hypothesis, however. Thus, I provided evidence for its validity. You could even say that the idea that the Bible is the Word of God is a scientifically-valid theory.

SOLUTIONS TO THE STUDY GUIDE FOR MODULE #3

1. a. <u>Experimental variable</u> - An aspect of an experiment which changes during the course of the experiment

b. <u>Control (of an experiment)</u> - The variable or part of the experiment to which all others can be compared

c. <u>Blind studies</u> - Experiments in which the participants do not know whether or not they are a part of the control group

d. <u>Double-blind studies</u> - Experiments in which neither the participants nor the people analyzing the results know who is in the control group

2. <u>An experimental variable is good when you are using it to learn something from the experiment. An experimental variable should be reduced or eliminated when it affects the results of the experiment but you do not learn anything from it</u>. In Experiment 3.2, for example, the type of "motor" in the "boat" was an experimental variable. It was a good variable, though, because you were using it to learn what kind of "motor" would work. The other experimental variables should have been reduced or eliminated, because they might have affected the results of the experiment but nothing would be learned from them.

3. <u>The control is the shirt that is being washed with no laundry detergent at all</u>. It is possible that all of the detergents are so bad that they have no real effect on the cleanness of the shirts. The only way to tell would be to compare it to a shirt that was washed in no detergent.

4. <u>The experimental variable that can be used to learn something from the experiment is the type of detergent used</u>.

5. There are at least four unwanted experimental variables. First, <u>the washers are different</u>. It is possible that some clean clothes better than others. This affects the results of the experiment, because you will not know whether the difference in cleanliness is due to the washer or the detergent. In addition, the <u>water can be at different temperatures</u>, which will affect the outcome. Also, <u>the shirts are different</u>. Some fabrics are easier to clean than others. Finally, <u>the amount of grass stain will be different in each shirt</u>, because there is no way to stain shirts equally.

6. <u>The experimental variable of the washers can be reduced by making sure all washers are the same brand and model, and by making sure they are all relatively new</u>. This will reduce the difference between the effectiveness of each washer. <u>You can reduce the differences in water temperature by monitoring the temperature of the water as it enters each washer and making adjustments to keep the temperature the same</u>. <u>The experimental variable of the shirts can be reduced by making sure they are all from the same manufacturer, the same style, and the same fabric</u>. That way, they are as close to identical as possible. Finally, <u>the experimental variable of the amount of stain</u> can be reduced by examining each stain carefully and trying to make sure they are as identical as possible.

7. <u>The data being collected are subjective</u>. Think about it. Each person's definition of "clean" is different. Also, the shirts are being examined by eye. This makes it hard to say exactly how much stain is left on each shirt. If you could chemically examine each shirt and determine precisely how much grass stain was left after washing, *that* would be an objective measurement. However, to have someone just look at a shirt and decide whether or not it is cleaner than another shirt is subjective.

8. The needle floats because of <u>surface tension</u>.

9. <u>Soap reduces the surface tension of water</u>.

10. <u>The liquid must have a larger surface tension than water</u>, because the needle floats more easily on this liquid than on water.

11. <u>You should give half of the volunteers the fat-free potato chips and the other half should get potato chips that have been on the market for years and seem to have no problems associated with them</u>. The latter group is the control. It's not enough to have the control group eat no potato chips, because the problem might just be with people eating *any* potato chips, not just the fat-free kind. <u>The volunteers then can keep a log (or you could observe them) for the next 4 hours to see if any stomach cramps occur</u>. If more cramps occur in the group that ate the fat-free chips than what occurred in the control group, then the allegations could be true. <u>This should definitely be a double-blind study</u>. If the volunteers knew which chip they were getting, it could bias them and they might imagine stomach cramps when, in fact, they had none. Also, comparing how two groups of people feel after eating is subjective. There is no way to get hard numbers from such a study. Thus, the person analyzing the data needs to be blind as well.

12. <u>This should be a blind study</u>. If the students knew whether or not they got the herb, it might influence how they take the test. However, since the data being collected is objective (measurable numbers), there is no reason for the person analyzing the data to not know who is in the control group and who isn't. There is no way he or she can bias the numbers.

13. <u>The study should be neither blind nor double-blind</u>. The experimental subjects are plants. They cannot "know" whether they are a part of the control or not. Thus, the whole idea of a blind study is kind of irrelevant. Also, the data being collected (the weight of the crop) is objective. The farmer can't bias the result, so there is no need for the farmer to be blind.

14. <u>This should be a double-blind study</u>. If the students know whether or not they are in the control group, it might influence how they behave. For example, the homeschooled students might be on their best behavior so as to give the researcher good results for homeschoolers. In the same way, the researcher's observations are subjective. There is no way to precisely measure how well one child interacts with another. It will depend heavily on the researcher's preconceived notions. Thus, the researcher must be as unbiased as possible and should therefore not know who is in the control group and who is not. In case you are interested, double-blind studies like this have been done, and the results are interesting. In general, such studies indicate the following observable differences:

a. Homeschooled students have fewer behavior problems than do publicly-schooled students.
b. Homeschooled students tend to be more openly friendly with those they have never met as compared to publicly-schooled students.
c. Homeschooled students tend to incorporate all ages, both genders, and all races in their games. Publicly-schooled students tend to play with those who are the same race, age, and gender.

15. <u>This should be a blind study, but it need not be double-blind</u>. The study needs to be blind because telling a person that he or she is on the "real" drug might affect his or her behavior. For example, a person might eat more than he or she usually does, assuming that he or she is "protected" from weight gain. Thus, everyone must be blind as to whether or they are in the control group. However, there is no need for the researcher to be blind, because the data are completely objective. The people each step on a scale and get weighed. The data, then, are a series of numbers that cannot be affected by the researcher's preconceived notions. Thus, whether or not he knows who is in the control group cannot affect the outcome of the experiment.

16. If there is no weight on the spring, it will not stretch. Thus, the length of the spring when it is not stretched out will be the length when no weight is hung on it. That means the value of the x-axis is zero. The dot that corresponds to an x-axis value of zero is halfway between the 4 and the 6 on the graph. Thus, the length of the spring with no weight on it is <u>5 inches</u>. If your answer is not exactly 5, that's okay. Somewhere between 4 and 6 is fine. Since you are reading from a graph, you do not exactly know where between 4 and 6 the dot is on the y-axis. If you look closely, however, you will see it is halfway between 4 and 6, which means 5.

17. If the spring is stretched to 8 inches, then the dot will be at 8 on the vertical axis. If you look below, I have drawn a line across the graph for a y-axis value of 8:

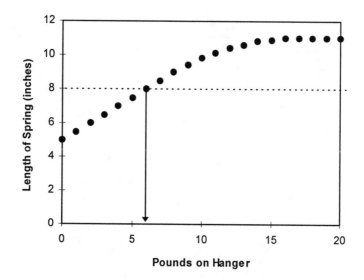

The dot that is at a y-value of 8 is just past the x-value of 5 (note the arrow). I would say it corresponds to <u>6 pounds</u>. An answer of 7 would be fine, because you cannot tell exactly. You just know it's between 5 and 10, but much closer to 5 than 10.

18. Notice how the y-values stop increasing after about 15 pounds. After that, no matter how many more pounds are put on the spring, it no longer stretches. Thus, <u>the spring stops stretching after about 15 pounds</u>. Once again, your number could be 14 or 16.

19. If the graphs all have the same shape, that means all springs stopped stretching after a certain number of pounds were placed on them. Thus, <u>springs stretch in response to a pull, but there is some maximum strength at which they simply no longer stretch</u>.

SOLUTIONS TO THE STUDY GUIDE FOR MODULE #4

1. a. <u>Simple machine</u> - A device that reduces the amount of force needed to perform a task or changes the direction of a force

b. <u>Force</u> - A push or pull that changes the motion of an object

c. <u>Mechanical advantage</u> - The amount by which force or motion is magnified in a simple machine

d. <u>Diameter</u> - The length of a straight line that travels from one side of a circle to another, while passing through the center of a circle

e. <u>Circumference</u> - The distance around a circle, equal to 3.1416 times the circle's diameter

2. <u>Applied science differs from science in motive. In applied science, the goal is to make something better. In science, the goal is simply to learn.</u>

3. <u>Technology can result from accident, science, or applied science.</u>

4. <u>Experiments (a) and (c) are applied science experiments, because the goal is to make something better. Experiments (b) and (d) are science experiments, because the goal is to learn something</u>. Even though the knowledge gained from experiments (b) and (d) might be useful, that's not the primary goal. Since the primary goal is knowledge, they are science experiments.

5. <u>Items (b) and (c) are technology.</u> They are not machines, but a vaccination for animals is something that makes life better, and a new diet for dogs does the same. Items (a) and (d) are simply pieces of information. They may be useful, but by themselves, they do not make life any better.

6. <u>The lever, the pulley, the wheel and axle, the inclined plane, the wedge, and the screw.</u>

7. <u>The inclined plane and the wedge look identical.</u>

8. For levers, the mechanical advantage equation (you have to memorize it) is:

Mechanical advantage = (distance from fulcrum to effort) ÷ (distance from fulcrum to resistance)

Mechanical advantage = 40 ÷ 10 = <u>4</u>

9. <u>The mechanical advantage means that the effort is magnified by 4</u>. It is impossible for the lever described in the problem to be a third-class lever, so the mechanical advantage definitely magnifies the effort, not the speed. However, if the student includes that in the answer, it is fine.

10. In a shovel, the part that does not move is the handle. You hold onto the handle with one hand, then you lift the middle of the shovel with the other. Thus, the effort is in the middle of the shovel. The resistance is in the shovel's head. Thus, the fulcrum is at one end, and the effort is between the fulcrum and the resistance. This is a <u>third-class lever</u>.

11. In a see-saw, the middle does not move. One child is the resistance and the other is the effort. Thus, the fulcrum is between the effort and the resistance. This is a <u>first-class lever</u>.

12. The mechanical advantage of a wheel and axle is given by:

Mechanical advantage = (diameter of the wheel) ÷ (diameter of the axle)

Mechanical advantage = 15 ÷ 3 = <u>5</u>

13. When the wheel is turned, a wheel and axle magnifies effort. Thus, <u>the applied force will be magnified 5 times</u>.

14. When the axle is turned, speed is magnified. Thus, <u>the wheel will move at 5 times the speed of the axle</u>.

15. The mechanical advantage of a multiple-pulley system is simply equal to the number of pulleys that work together. Thus, the mechanical advantage is <u>6</u>.

16. The mechanical advantage allows you to use less force when you lift, but you "pay" for that by having to pull that much more rope. If the person wants to lift the load 1 foot and the mechanical advantage is 6, the person will need to pull <u>6 feet</u> of rope.

17. The mechanical advantage of an inclined plane is given by:

Mechanical advantage = (length of slope) ÷ (height)

Mechanical advantage = 6 ÷ 2 = <u>3</u>

18. The mechanical advantage equation for a wedge is the same as it is for the inclined plane. Since both have the same dimensions, the mechanical advantage is the same: <u>3</u>.

19. The mechanical advantage equation for a screw depends on the circumference. Thus, we need to calculate that first:

Circumference = 3.1416 x (diameter)

Circumference = 3.1416 x 2 = 6.2832

Now we can use the mechanical advantage equation for a screw:

Mechanical advantage = (circumference) ÷ pitch

Mechanical advantage = 6.2832 ÷ 0.1 = <u>62.832</u>

20. <u>You should get a fatter screwdriver</u>. The mechanical advantage of a screw/screwdriver combination is dependent on the circumference of the screwdriver. Thus, the fatter the screwdriver, the better!

SOLUTIONS TO THE STUDY GUIDE FOR MODULE #5

1. a. <u>Life science</u> – A term that encompasses all scientific pursuits related to living organisms

b. <u>Archaeology</u> – The study of past human life as revealed by preserved relics

c. <u>Artifact</u> - Objects made by humans such as tools, weapons, containers, etc.

d. <u>Geology</u> – The study of earth's history as revealed in the rocks that make up the earth

e. <u>Paleontology</u> – The study of life's history as revealed in the preserved remains of once-living plants and animals

f. <u>Aristotle's dictum</u> - The benefit of the doubt is to be given to the document itself, not assigned by the critic to himself.

g. <u>Known age</u> - The age of an artifact as determined by a date printed on it or a reference to the artifact in a work of history

h. <u>Dendrochronology</u> - The process of counting tree rings to determine the age of a tree

i. <u>Radiometric dating</u> - Using a radioactive process to determine the age of an item

j. <u>Absolute age</u> – The calculated age of an artifact when a specific dating method is used to determine when the artifact was made

k. <u>The Principle of Superposition</u> – When artifacts are found in rock or earth that is layered, the deeper layers hold the older artifacts.

2. <u>You would use paleontology</u>, because archaeology concentrates on human life.

3. <u>The internal test, the external test, and the bibliographic test.</u>

4. <u>The internal test makes sure that the document does not contradict itself. The external test makes certain that the document does not contradict other known historical or archaeological facts. The bibliographic test makes certain that the document we have today is essentially the same as the original.</u>

5. <u>Aristotle's dictum is used in the internal test. We must use it because what seems to be a contradiction in a document might not be a contradiction. It might just be our inability to understand the language in which the document was written.</u>

6. <u>Often those who are making the copy or those who are ordering the copy to be made will order changes to be made as well.</u> Kings have done this in an effort to make themselves or their

ancestors look better in history. Religious groups have been known to do this to make themselves look more important or to make their view look "right."

7. <u>First, there should be a small time period between when the original was written and when the first available copy was made</u>. This reduces the chance for changes being made and reduces the number of errors that would be committed during the copy process. <u>Second, there must be a lot of different copies from a lot of different sources</u>. If all of the copies agree with one another, then we know that a single copier did not make drastic changes.

8. <u>No</u>. The Bible passes the internal test as well as any document of its time.

9. <u>Yes</u>. Because of the difficulty of translating ancient languages, there are some difficult passages. All documents of history have such passages, however.

10. This is a translation problem. The verb "hear" used in Acts 9:7 simply means that the men heard sounds. The verb "hear" used in Acts 22:9 requires that the hearer must actually understand intelligible language. <u>These verses are really complementary, then. The first tells us that the men heard SOUNDS, but the second tells us that the men could not understand those sounds</u>.

11. <u>One of the genealogies traces Mary's line while the other traces Joseph's line</u>.

12. We can say this because <u>no other work has had so much archaeological evidence compared to it</u>. The Bible has been tested by archaeology more than any other documents of history, and it passes with flying colors!

13. <u>Sometimes, it turns out that archaeology is wrong</u>, so you cannot discount the validity of a document if archaeology does not fully agree with it. Remember, several archaeologists thought that the Bible was wrong on several occasions. It turns out that it was the archaeologists who were wrong, not the Bible!

14. <u>The New Testament has significantly shorter time spans between original and copy as compared to any other work of the same time period. It also has thousands more supporting documents than any other document of its time</u>.

15. <u>Yes, the Old Testament passes the bibliographic test just as well as any other document of its time</u>.

16. <u>The age is absolute</u>, because a dating method was used to determine it.

17. <u>The coffin has a known age</u>, because it is referenced in a document of history.

18. <u>NO. Absolute does not mean certain</u>. Even the most accurate dating method has error in it, and some dating methods can be very unreliable.

19. <u>Master tree ring patterns help the archaeologist determine the age of logs found in ruins.</u> Master tree ring patterns are cataloged for each region of the world, and they correspond to weather patterns that have already been dated. If an archaeologist finds a master tree ring pattern on a log, he or she knows when that tree ring pattern was formed and can use that to determine the age of the log.

20. <u>The Principle of Superposition assumes that in rock or soil that is layered, the layers were formed one at a time.</u> This is not necessarily true.

21. <u>He can conclude that the city he found was built before 2500 B.C.</u> Assuming the Principle of Superposition is true, the lower layers of rock are older than the upper layers. Since he found this city in a lower layer of rock, it must be older than the city that was discovered in the upper layer of rock.

22. <u>There are many seemingly unrelated cultures that all have a worldwide flood tale.</u> If the flood did not really occur, then you have to assume that they all made up the tale independently, because many of the cultures had no contact with one another until well after the tales were written down.

SOLUTIONS TO THE STUDY GUIDE FOR MODULE #6

1. a. <u>Catastrophism</u> - The view that most of earth's geological features are the result of large-scale catastrophes such as floods, volcanic eruptions, etc.

b. <u>Uniformitarianism</u> - The view that most of earth's geological features are the result of slow, gradual processes that have been at work for millions or even billions of years

c. <u>Humus</u> - The decayed remains of once-living creatures

d. <u>Minerals</u> - Inorganic crystalline substances found naturally in the earth

e. <u>Weathering</u> - The process by which rocks are broken down by the forces of nature

f. <u>Erosion</u> - The process by which sediments and rock fragments are carried away by wind or rain

g. <u>Unconformity</u> - A surface of erosion that separates one layer of rock from another

2. <u>The uniformitarian hypothesis</u> requires that the earth be billions of years old, because it assumes that the geological features of the earth took millions and billions of years to form. Catastrophism is more flexible. It can accommodate a young earth or an earth that is billions of years old.

3. The three basic types of rock are <u>igneous, metamorphic, and sedimentary</u>.

4. <u>Igneous rock is the result of molten rock which cools and solidifies. Sedimentary rock is formed when sediments fuse together. Metamorphic rock is formed when either sedimentary or igneous rocks are exposed to extreme pressure.</u>

5. Most sedimentary rock has been laid down by <u>water</u>.

6. This is <u>physical weathering</u>. The small chips of rock that are broken off are just miniature versions of the original rock. No change in composition has occurred.

7. This is <u>chemical weathering</u>. The limestone forms a gas. That changes the composition of what's left.

8. You expect the most erosion from the <u>quickly-flowing river</u>. Remember, your experiment showed that the faster the water moves, the more erosion occurs.

9. <u>The barren hillside will experience the most erosion</u>. Remember, your experiment showed that plants reduce the effects of erosion.

10. Many sediments carried by a river are deposited in <u>the river's delta</u>.

11. Underground caverns are formed by <u>erosion caused by groundwater</u>.

12. <u>Stalactites form on the ceiling of a cavern, while stalagmites form on the floor of a cavern</u>.

13. <u>Stalactites and stalagmites are formed when groundwater seeps through the ceiling of a cavern. As the drop forms and falls to the floor of the cavern, it might deposit sediments on the ceiling or floor. As those sediments pile up, stalactites and stalagmites are formed</u>.

14. <u>C</u>

15. <u>A and E</u>. Remember, igneous rock is formed from magma. In A, the magma came from a volcanic eruption. In E, it broke into the layers of sedimentary rock.

16. <u>D</u>

17. <u>B: It is an angular unconformity</u>.

18. <u>E</u>

SOLUTIONS TO THE STUDY GUIDE FOR MODULE #7

1. a. <u>Petrifaction</u> - The conversion of organic material into stone

b. <u>Resin</u> – A sticky, liquid substance that usually hardens when exposed to air

c. <u>Extinct</u> - A term applied to a species of plant or animal that was once living but now is not

2. <u>The most likely thing that will happen to the remains of a dead plant or animal is that they will decompose.</u> Fossilization is a rare exception to this general rule.

3. <u>A fossil mold forms first.</u> If a cast forms, it forms later when the mold is filled with sediment.

4. <u>The remains of a plant or animal are encased in sediment and the sediment eventually hardens into rock. As the remains of the plant or animal disintegrate, a hole is left in the rock, in the shape of the original remains. That is the mold. The mold might fill up with sediment later and, when the sediment hardens, it forms a cast.</u>

5. <u>Petrifaction requires water that has a lot of minerals in it.</u>

6. <u>Petrified fossils have more information than fossil casts because fossil casts retain only the shape and outer details of the fossil. When a fossil is petrified, its components are replaced with minerals. This means the entire fossil is preserved, which gives us more information than just the shape and outer details of the fossil.</u>

7. <u>A carbonate residue still has a film residue of the original creature. In an impression, the film is completely gone, leaving only an outline that looks like it was etched in stone.</u>

8. <u>Both carbonate residues and impressions are formed by the same process. They also leave an outline of the creature.</u>

9. Fossils encased in amber or ice do not decompose as quickly as other fossils. <u>Thus, tissue and other soft parts which usually are not preserved tend to be preserved very well.</u>

10. The four general features of the fossil record are:

1. <u>Fossils are usually found in sedimentary rock. Since most sedimentary rock is laid down by water, it follows that most fossils were laid down by water as well.</u>
2. <u>The vast majority of the fossil record is made up of clams and other hard-shelled creatures. Most of the remaining fossils are of water-dwelling creatures and insects. Only a tiny, tiny fraction of the fossils we find are of plants, reptiles, and mammals.</u>
3. <u>Many of the fossils we find are of plants and animals which are still alive today. Some of the fossils we find are of plants and animals which are now extinct.</u>
4. <u>The fossils found in one layer of stratified rock can be considerably different than the fossils found in another layer of the same stratified rock.</u>

11. Clams and other hard-shelled animals make up 95% of the fossil record.

12. Approximately 1,000 species have gone extinct over the last 400 years. This is a stark contrast to the 10,000 species which some "environmentalists" claim go extinct *each year*!

13. A trilobite is a creature that lived in the water and was covered in a hard outer covering. Typically, trilobites lived at the bottom of the ocean. They are now assumed to be extinct.

14. A placoderm is a kind of fish. It was much like the fish we see today, but its head was covered with hard plates rather than scales. Placoderms are considered to be extinct.

15. According to uniformitarians, sediments are laid down slowly over millions of years. Eventually, conditions change and the sediments harden to form rocks. The conditions during which the sediments were laid down determine the type of sediment, which in turn determines the kind of rock formed.

16. According to catastrophists, most of the sedimentary rocks we see today were formed in Noah's flood. The depth, speed, and direction of the flood waters determined what kind of sediments were laid down, which in turn determined the type of rock formed.

17. According to uniformitarians, each layer of rock represents a period of earth's history. Thus, the different fossils found in different layers result from the fact that different plants and animals existed at different times in any given region.

18. According to catastrophists, most of the sedimentary rock we see today is the result of the flood. Thus, the depth, speed, and direction of the flood waters determined where the fossils being preserved came from. Thus, the different fossils in different layers are the result of the fact that different kinds of fossils were trapped and preserved during different stages of the flood.

19. Uniformitarians must speculate how millions of year of time affect the processes that we see working today. At best, we have viewed how these processes work over a few thousand years. The effect that millions of years will have on the processes can only be speculated.

20. Catastrophists must speculate about the nature of Noah's flood. The speculation is aided by the observation of local catastrophes. Nevertheless, Noah's flood would have been much different than a local catastrophe, so the details of the flood can only be speculated.

SOLUTIONS TO THE STUDY GUIDE FOR MODULE #8

1. a. <u>Index fossils</u> - A fossil considered to represent a certain period in earth's past

b. <u>Geological column</u> - A theoretical picture in which layers of rock from around the world are meshed together into a single, unbroken record of earth's past

c. <u>The Theory of Evolution</u> - A theory which states that all of life on this earth has a single, common ancestor that existed a long time ago

2. <u>Index fossils are used by geologists to determine what time period a layer of rock represents.</u> If a geologist finds index fossils for the Cambrian time period in a layer of rock, for example, the geologist says that the layer of rock was laid down during Cambrian times.

3. <u>The geological column is constructed by comparing layers of rock found in various parts of the earth. Using index fossils and the Principle of Superposition, geologists order the layers into one, big column which represents all of earth's geological history.</u>

4. <u>No.</u> The geological column is not something that exists somewhere in the world. It is a construct based on uniformitarian assumptions.

5. Since trilobites are lower on the geological column, uniformitarians assume that <u>trilobites existed on earth before fish.</u>

6. When only algae are found in a layer of rock, the geological column says that the rock is assumed to be Precambrian, which is assumed to be <u>570 millions years old or older.</u>

7. According to the geological column, fish came before mammals. This means the layer with only fish must be the older layer. Thus, <u>the layer with fish should be on the bottom and the layer with mammals should be on top.</u>

8. <u>The geological column is viewed as evidence for evolution because it indicates that early in earth's history, there were only simple life forms. As time went on, the geological column indicates that more and more complex life forms started to appear.</u> This is exactly what the Theory of Evolution says.

9. <u>The geological column is not really evidence for evolution because it is not real. Since it is constructed with assumptions, the evidence is only good if the assumptions are valid.</u>

10. The data from Mt. St. Helens indicate that <u>stratified rock can form in less than a day.</u>

11. <u>No, you should not make that assumption.</u> The only time we have actually observed a canyon form, it formed as the result of a catastrophe and then later the river formed. Thus, the data from Mt. St. Helens tells us that canyons cause rivers, not that rivers cause canyons.

12. <u>On a geological scale, the Mt. St. Helens catastrophe was rather minor. This tells us that a major catastrophe would most likely result in larger deposits of stratified rocks and larger canyons.</u>

13. The Mt. St. Helens eruption had <u>mudflows</u>. These mudflows caused most of the geological features that have been studied, and a major flood would also have such mudflows.

14. <u>The Cumberland Bone Cave is a fossil graveyard that contains many fossils from several different climates. This is important because it is excellent evidence for a worldwide flood and is a major problem for the uniformitarian viewpoint because it is not something that can be well explained in terms of uniformitarian assumptions.</u>

15. <u>No</u>. Fossilized hats, pliers, and waterwheels tell us that fossils can form rapidly.

16. <u>A paraconformity is an unconformity that does not really exist in a geological formation but uniformitarians believe must exist because of the fossils found in the formation.</u>

17. The text discusses the following problems - you need only list 4 of them:

a. <u>There are too many fossils in the fossil record.</u>

b. <u>Fossil graveyards with fossils from many different climates are hard to understand in the uniformitarian view.</u>

c. <u>Index fossils are called into question by the many creatures we once thought were extinct but now know are not.</u>

d. <u>Uniformitarians must assume the existence of paraconformities.</u>

e. <u>Uniformitarians must believe that evolution occurred, and there is no evidence for evolution.</u> In fact, the fossil record provides evidence that each plant and animal was created by God.

18. <u>Catastrophists have offered no good explanation for the existence of unconformities between rock layers laid down by the flood. They also cannot explain certain fossil structures that look like they were formed under "normal" living conditions which would not exist during the flood. Finally, they have not yet explained the enormous amount of limestone that exists all over the world in terms of Noah's flood.</u>

19. <u>The fossil record contains no fossils that are undeniable intermediate links</u>. If evolution occurred, there should be millions of such fossils.

20. <u>The fossil record contains no fossils that are undeniable intermediate links</u>. This is exactly what you would expect if God created each plant and animal individually.

SOLUTIONS TO THE STUDYGUIDE FOR MODULE #9

1. a. <u>Atom</u> - The smallest stable unit of matter in Creation

b. <u>Molecule</u> - Two or more atoms linked together to make a substance with unique properties

c. <u>Photosynthesis</u> - The process by which a plant uses the energy of sunlight and certain chemicals to produce its own food

d. <u>Metabolism</u> - The process by which a living organism takes energy from its surroundings and uses it to sustain itself, develop, and grow

e. <u>Receptors</u> - Special structures or chemicals that allow living organisms to sense the conditions of their surroundings

f. <u>Cell</u> - The smallest unit of life in Creation

2.
*   <u>All life forms contain DNA.</u>

*   <u>All life forms have a method by which they take energy from the surroundings and convert it into energy that helps them live.</u>

*   <u>All life forms can sense changes in their surroundings and respond to those changes.</u>

*   <u>All life forms reproduce.</u>

3. DNA provides the <u>information</u> necessary to turn lifeless chemicals into a living organism.

4. It is <u>big</u>. In fact, DNA is one of the biggest molecules in Creation.

5. <u>DNA is significantly more efficient at information storage than the best computer human science can make.</u>

6. a. <u>The nucleotides</u> store the information. Remember, the information of a living organism is stored as a sequence of nucleotides.

b. <u>The backbone</u> forms long ribbons that twist to make the double helix structure.

7. <u>Thymine links to adenine, and guanine links to cytosine.</u>

8. The relationship in 7 allows you to determine the other half of the DNA, because only adenine and thymine can link up. Similarly, only cytosine and guanine can link up.

<u>guanine, cytosine, thymine, cytosine, adenine, adenine</u>

9. The relationship in 7 allows you to determine the other half of the DNA, because only adenine and thymine can link up. Similarly, only cytosine and guanine can link up.

<u>adenine, cytosine, guanine, thymine, adenine, cytosine</u>

10. Plants use photosynthesis to make their own food, <u>glucose</u>.

11. Plants store food as <u>starch</u>.

12. A biosphere is <u>a living system that is isolated from its surroundings but still continues to live and grow</u>.

13. Metabolism requires food and <u>oxygen</u>.

14. Metabolism usually produces energy, <u>carbon dioxide, and water</u>.

15. <u>The organism will not be able to sense and respond to change</u>.

16. <u>The anchovy parents will have many more babies</u>, because so many anchovies get eaten that many must be born to "replace" them.

17. <u>The cat is still alive</u>, because its cells can reproduce.

18. <u>No, it is not</u>. The speed at which the human population is increasing has been slowing every year. Also, there is more food per person today than ever before. Finally, the cost of raw materials is lower than ever. Thus, there are no indicators which point to trouble.

19. See Figure 9.3

a. <u>organelles</u>
b. <u>cytoplasm</u>
c. <u>nucleus</u>
d. <u>membrane</u>

20. DNA is stored in the <u>nucleus</u>.

21. There are <u>three</u> basic kinds of cells: plant cells, animal cells, and cells from bacteria.

22. <u>The scientist will see two basic kinds of cells</u>. The leaf cell will be a plant cell, and the mouse cell and cat cells will both be animal cells.

SOLUTIONS TO THE STUDY GUIDE FOR MODULE #10

1. a. <u>Prokaryotic cell</u> - A cell that has no organelles

b. <u>Eukaryotic cell</u> - A cell with organelles

c. <u>Pathogen</u> - An organism that causes disease

d. <u>Decomposers</u> - Organisms that break down the dead remains of other organisms

e. <u>Vegetative reproduction</u> - The process by which the stem of a plant can form new roots and develop into a mature plant

2. The five kingdoms are: <u>Monera, Protista, Fungi, Plantae, and Animalia.</u>

3. Since it is made up of several eukaryotic cells, it is not in Kingdom Monera. Since it eats dead organisms, it is probably a decomposer. Thus, it is in kingdom <u>Fungi</u>.

4. Since it is made up of several eukaryotic cells, it is not in kingdom Monera. Since it makes its own food, it is probably a plant. However, now that you know about algae, you have to be a bit concerned, because it might be algae and thus belong to kingdom Protista. However, algae does not have specialized structures, so this is not algae. It therefore must be in kingdom <u>Plantae</u>.

5. If it is a single, prokaryotic cell, it must be in kingdom <u>Monera</u>. The part about eating dead organisms is just there to fool you. Regardless of its eating habits, because it is a prokaryotic cell, it can *only* be a part of kingdom Monera.

6. Since it is made up of several eukaryotic cells and makes its own food, it is either algae or a plant. Since it has no specialized structures, it is algae. This puts it in kingdom <u>Protista</u>.

7. If it is a single, eukaryotic cell, it is most likely in kingdom <u>Protista</u>. If it ate only dead organisms, it could be a single-celled fungus. However, the problem says it eats other, living organisms. That means it is not in kingdom Fungi.

8. Since it is made up of several eukaryotic cells, it is not in kingdom Monera. Since it does not make its own food, it is not a plant and thus not in Plantae. Since it is more than one cell and is not algae, then it is not in kingdom Protista. It eats living plants, so it is not in kingdom Fungi, either. Thus, it must be in kingdom <u>Animalia</u>.

9. <u>Bacteria must have water to survive. Dehydrated food has almost all of the water removed, so bacteria cannot survive to grow and reproduce.</u>

10. <u>The presence of salt reduces the growth and reproduction of bacteria.</u> Thus, salt protects meat from contamination by bacteria.

11. <u>Bacteria can be introduced onto food or liquid by dust particles in the air. Covering the food keeps the dust particles off, preventing the addition of new bacteria onto the food</u>.

12. If it can move on its own, it is a <u>protozoa</u>.

13. <u>Yes</u>, there are many pathogenic organisms in kingdom Protista. The text gives an example.

14. <u>Decomposers recycle the dead matter in Creation. Without them, there would be no way that the materials in dead organisms could be used again by living organisms</u>. One reason Biosphere 2 failed was that it didn't have a large enough variety of decomposers.

15. <u>No, not all members of kingdom Fungi are made of several cells. Yeast are an example of single-cell fungi</u>.

16. <u>Like an iceberg, the visible part of a mushroom is actually a pretty small portion of the organism. Most of the organism lies unseen underground</u>.

17. The <u>mycelium</u> is actually the main part of the mushroom.

18. Only <u>plant cells</u> have cell walls and central vacuoles. It cannot be prokaryotic, because the central vacuole is an organelle!

19. <u>Turgor pressure is the pressure inside of a plant cell that is caused by the central vacuole pushing the cell contents against the cell wall. It allows plants to stand upright</u>.

20. Humans are members of kingdom <u>Animalia</u>.

SOLUTIONS TO THE STUDY GUIDE FOR MODULE #11

1 a. Axial skeleton – The portion of the skeleton that supports and protects the head, neck, and trunk

b. Appendicular skeleton – The portion of the skeleton that attaches to the axial skeleton and has the limbs attached to it

c. Exoskeleton – A body covering, typically made of a tough, flexible substance called chitin, that provides support and protection

d. Symbiosis – Two or more different organisms living together so that each benefits from the other

2. The human superstructure is made up of the skeleton, the muscles, and the skin.

3. One major difference is their appearance. Smooth muscles are smooth and unstriped while skeletal muscles are rough and striped. The other main difference is the way they operate. Skeletal muscles are voluntary (they are operated by consciously thinking about it), while smooth muscles are involuntary (they are operated unconsciously by the brain).

4. The cardiac muscle is in the heart. It is an involuntary muscle

5. The bone marrow produces blood cells.

6. Keratinization is a process that hardens living cells. It is used to make the outer layer of the epidermis, as well as hair and nails. Remember, keratinization kills cells.

7. Bones are made up of collagen and minerals. The collagen makes the bones flexible, while the minerals make them hard and strong.

8. Compact bone tissue is packed together tightly while spongy bone tissue has lots of space in between its fibers. Because of this, spongy bone tissue is lighter than compact bone tissue.

9. Bones are definitely alive. There are living cells imbedded in the bone. Thus, bone tissue is living tissue. That's why it can grow!

10. Vertebrates are animals with backbones. Invertebrates are animals without backbones. It is possible to be neither. If an organism is from *any* kingdom other than Animalia, it is neither an invertebrate nor a vertebrate.

11. According to their definitions, the arms belong to the appendicular skeleton, but the neck is a part of the axial skeleton. See Figure 11.2.

12. <u>An exoskeleton is a support structure that exists on the *outside* of an organism, while an endoskeleton is on the *inside* of the organism. Creatures with exoskeletons are called arthropods.</u>

13. In terms of range of motion, the hinge offers the least and the ball-and-socket offers the most. Thus, the order in increasing range of motion is: <u>hinge, saddle, ball-and-socket</u>. When there is a larger range of motion, the joint is less stable. Thus, the order in terms of stability is reversed: <u>ball-and-socket, saddle, hinge</u>.

14. <u>Ligaments tend to hold the bones of the joints together. Cartilage cushions the bones of the joints so that they do not rub painfully against each other.</u>

15. <u>Skeletal muscles end in tendons, and the tendons attach to the skeleton.</u>

16. <u>In order to raise the forearm, the biceps contracts while the triceps relaxes. To extend the forearm, the biceps relaxes while the triceps contracts.</u>

17. <u>The stomach is made of smooth muscle.</u> You know that because you do not have to think about working your stomach in order for it to do its job.

18. Plants' ability to grow towards the light is called <u>phototropism</u>.

19. Hair is used to <u>insulate and provide sensation</u>.

20. Sweat <u>cools the body down and also provides food for beneficial bacteria and fungi which live on your skin</u>.

21. Skin cells constantly fall off your body because <u>the cells on the outer layer are dead</u>.

22. The sebaceous glands produce <u>oil</u>. This oil <u>softens the skin and hair and also makes it hard for certain bacteria to attach themselves to your skin</u>.

23. Animals whose skin produces hair are typically mammals. If the skin produces feathers, the animal is a bird. Scales indicate a reptile, and breathing through the skin indicates an amphibian. The last classification is not ironclad. Worms, for example also breathe through their skin, but they are not amphibians.

   a. <u>Mammal</u>
   b. <u>Amphibian</u>
   c. <u>Reptile</u>
   d. <u>Bird</u>

SOLUTIONS TO THE STUDY GUIDE FOR MODULE #12

1. a. <u>Producers</u> - Organisms that produce their own food

b. <u>Consumers</u> - Organisms that eat living producers and/or other consumers for food

c. <u>Herbivore</u> - A consumer that eats producers exclusively

d. <u>Carnivore</u> - A consumer that eats only other consumers

e. <u>Omnivore</u> - A consumer that eats both plants and other consumers

f. <u>Basal metabolic rate</u> - The minimum amount of energy required by the body every day

2. The energy in living organisms originates in the <u>sun</u>.

3. a. A mushroom is in kingdom Fungi and is therefore a <u>decomposer</u>.

b. An evergreen bush is a plant and is therefore a <u>producer</u>.

c. A worm eats other things (it is certainly not a plant!), thus it is a <u>consumer</u>.

d. In Module #10, you learned that algae are the most important source of photosynthesis on the planet. Thus, they are <u>producers</u>.

4. Food is converted to energy via the process of <u>combustion</u>.

5. <u>Combustion requires oxygen</u>, as well as something to burn, like wood or monosaccharides.

6. <u>Combustion produces energy, carbon dioxide, and water</u>.

7. The three macronutrients are <u>carbohydrates, fats, and proteins</u>.

8. The main thing that the macronutrients provide is <u>energy</u>.

9. We need to eat a lot more <u>macronutrients</u>.

10. Disaccharides are made of 2 monosaccharides linked together, while polysaccharides are made of *several* monosaccharides linked together. Thus, <u>a polysaccharide is the largest</u>.

11. <u>Glucose is a monosaccharide</u>. Most carbohydrates contain a lot of glucose.

12. Fats come in two types: <u>saturated fats</u> and <u>unsaturated fats</u>. You can distinguish them by looking at them while they are at room temperature. <u>Saturated fats are solid at room temperature, while unsaturated fats are liquid</u>.

13. Proteins are made of long strings of <u>amino acids</u>.

14. The body prefers to burn <u>carbohydrates, then fats, and then proteins (or amino acids)</u>.

15. <u>If your cells do not have enough amino acids, the amino acids from the proteins you eat are shipped to your cells so that your cells can make the proteins they need. If your cells have plenty of amino acids, the amino acids from the proteins you eat are converted into carbohydrates or fats</u>.

16. <u>Your cells must make proteins by linking together amino acids. There are several amino acids which your body cannot make. Thus, you must get them from food. Without those amino acids, your cells will not be able to make the proteins they need to make</u>. Animal proteins have these amino acids in plentiful supply. Plant proteins rarely have all of them. Thus, people who eat only plants must get a wide variety of plant proteins to make sure they get those amino acids.

17. <u>Endothermic animals have higher BMRs</u>. The BMR tracks the minimum amount of energy needed to survive. Both endothermic and ectothermic animals have involuntary muscles that need energy, but only endothermic animals expend energy to keep their internal temperatures high.

18. <u>Ectothermic animals are slower on cold days</u>. Since their body temperatures are not held constant, the colder days reduce the speed at which the chemical reactions can occur in their bodies. This makes them sluggish.

19. <u>Calories are a measure of energy</u>. They can be used to measure how much energy is in food, or how much energy is expended by a living organism.

20. <u>The second man is less active during the day</u>. Since they both burn about the same number of calories while sleeping, they both have essentially the same BMR. Remember, BMR is the minimum amount of energy you need. The amount of energy you burn when you sleep is pretty minimal. If the first needs more food, he must be burning more energy during the day, when both men are active.

21. <u>No, you cannot. It might be that Jean's BMR is simply much higher than Wanda's</u>. They can each be very active, but their BMR is a major factor in determining how much food they need.

22. In general, the smaller the mammal the higher the BMR, because the more energy the animal has to spend keeping its internal temperature high. Thus, the <u>pig</u> has the higher BMR.

23. <u>Combustion of food takes place in the cell</u>.

24. The mitochondrion is called the "powerhouse" of the cell because <u>the majority of energy in the combustion process is released in step 3, which takes place in the mitochondrion</u>.

SOLUTIONS TO THE STUDY GUIDE FOR MODULE #13

1. a. <u>Digestion</u> - The process by which an organism breaks down its food into small units which can be absorbed

b. <u>Vitamin</u> - a chemical substance essential for the normal working of the human body

2. a. <u>tongue</u>   b. <u>larynx</u>   c. <u>trachea</u>   d. <u>pancreas</u>   e. <u>stomach</u>   f. <u>small intestine</u>

g. <u>rectum</u>   h. <u>salivary glands</u>   i. <u>pharynx</u>   j. <u>esophagus</u>   k. <u>lungs</u>   l. <u>liver</u>

m. <u>gall bladder</u>   n. <u>large intestine</u>   o. <u>appendix</u>   p. <u>anus</u>

3. a. <u>The tongue is a part of the digestive system and a part of the digestive tract. It moves the food around in the mouth to form the bolus.</u> It also provides a sense of taste, but that's more a part of its nervous system function, which you will learn about later.

b. <u>The larynx is not a part of the digestive system.</u>

c. <u>The trachea is not a part of the digestive system.</u>

d. <u>The pancreas is a part of the digestive system but not a part of the digestive tract. It has two main functions. It makes several digestive juices which are squirted into the small intestine as chyme passes through the pyloric sphincter. In addition, it also produces a base to neutralize the stomach acid.</u> The base is called sodium bicarbonate, but you do not have to know that.

e. <u>The stomach is a part of the digestive system and the digestive tract. It churns and mixes the food with gastric juices. The gastric juices contain stomach acid which destroys bacteria that might have been eaten with the food and helps dissolve the food. The gastric juices also contain some digestive chemicals that start the chemical digestion of the food. This turns the bolus of food into chyme.</u>

f. <u>The small intestine is a part of the digestive system and the digestive tract. It chemically digests the food and allows the nutrients to be absorbed through its lining.</u>

g. <u>The rectum is a part of the digestive system and the digestive tract. It pushes food out of the body through the anus.</u>

h. <u>The salivary glands are a part of the digestive system. Technically, they are not part of the digestive tract because food does not pass through them.</u> Don't worry if you got that wrong. <u>They put saliva in the mouth. The saliva partially digests the food, but it also lubricates the mouth and makes it easier for the tongue to form the bolus.</u>

i. The pharynx is a part of the digestive system and the digestive tract. It pushes food into the esophagus.

j. The esophagus is a part of the digestive system and the digestive tract. It pushes food down into the stomach.

k. The lungs are not a part of the digestive system.

l. The liver is a part of the digestive system but not a part of the digestive tract. It has many functions. The most important digestion-related functions are making bile, converting monosaccharides to glycogen for storage, breaking glycogen down when the body needs energy, storing amino acids, storing fats, and converting fats and amino acids into monosaccharides when the body needs energy. The liver has many other functions, but those are the ones I want you to remember.

m. The gall bladder is a part of the digestive system but not a part of the digestive tract. It concentrates bile and squirts the bile into the chyme as the chyme enters the small intestine. Bile is a chemical that aids in the digestion of fats.

n. The large intestine is a part of the digestive system and the digestive tract. It consolidates undigested food, absorbs water from it, and turns the resulting waste into feces. Bacteria in the large intestine also produce vitamin K.

o. The appendix is a part of the digestive system. It is not a part of the digestive tract. You might have thought that it was since it is attached to the large intestine, but it is not. It sort of "hangs" off the large intestine and food never passes through it. Its function is not known. Many think that it is a highly-specialized disease-fighting organ.

p. The anus is a part of the digestive system and the digestive tract. It is the opening through which feces exit.

4. The epiglottis covers the larynx when you swallow. It stays open when you breathe.

5. The digestive process would essentially stop in the small intestine, because the digestive chemicals in the small intestine cannot work in the presence of acid.

6. Vitamins A, D, E, and K are fat-soluble.

7. The fat-soluble vitamins are the most likely to build up to toxic levels, as the water-soluble vitamins can be ejected from the body through the urine.

8. Vitamins typically regulate the chemical processes in the body.

9. Vitamins D and K. Vitamin D is made from sunlight hitting the skin, and vitamin K is made by bacteria in the large intestine.

SOLUTIONS TO THE STUDY GUIDE FOR MODULE #14

1. a. Veins - Blood vessels that carry blood back to the heart

b. Arteries - Blood vessels that carry blood away from the heart

c. Capillaries - Tiny, thin-walled blood vessels that allow the exchange of gases and nutrients between the blood and cells

2. The lungs oxygenate the blood and allow the blood to get rid of carbon dioxide.

3. The heart pumps blood throughout the circulatory system.

4. The human heart has four chambers: the right atrium, left atrium, right ventricle, and left ventricle.

5. A four-chambered heart separates deoxygenated blood from oxygenated blood. This is much more efficient than other kinds of hearts.

6. Capillaries are the vessels which allow exchange of gases between the cells and the blood.

7. Deoxygenated blood comes into the heart through the vena cava and gets dumped into the right atrium.

8. After filling the right atrium, the deoxygenated blood is pushed into the right ventricle.

9. When deoxygenated blood leaves the heart, it heads for the lungs to become oxygenated.

10. Oxygenated blood that is returning to the heart fills the left atrium.

11. After entering the left atrium, the oxygenated blood is dumped into the left ventricle.

12. When oxygenated blood is leaving the heart, it is pumped out into the aorta.

13. Veins usually carry deoxygenated blood, because they are carrying it back to the heart to be pumped into the lungs. There are exceptions. The pulmonary vein, for example, carries oxygenated blood to the heart from the lungs. You need not know the name of the exception.

14. The red blood cells carry oxygen to the other cells of the body. The white blood cells fight disease-causing organisms, and the blood platelets aid the coagulation process when a blood vessel is injured.

15. More than half of your blood is blood plasma.

16. Hemoglobin is a protein that carries oxygen. It is found in the red blood cells.

17. Blood cells are produced in <u>bone marrow</u>.

18. <u>Alveoli are small sacks at the end of tiny bronchial tubes. Oxygenation of blood takes place in them. The blood also gives up waste products there.</u>

19. <u>Capillaries surround the alveoli</u>, since that's where the blood gets oxygen and releases wastes.

20. Bronchial tubes are found in the <u>lungs</u>. <u>Air</u> travels through them.

21. <u>nasal cavity, pharynx, larynx, trachea, bronchial tubes, alveoli</u>

22. Vocal cords are found in the <u>larynx</u>.

23. <u>The amount of air that passes over the vocal cords controls the volume of the sound, and the tightness of the vocal cords determines the pitch.</u>

24. The xylem transport water up the plant. The phloem transport food down the plant. Thus, this sample was taken from the plant's <u>xylem</u>.

SOLUTIONS TO THE STUDY GUIDE FOR MODULE #15

1. a. <u>Gland</u> - A group of cells that prepare and release a chemical for use by the body

b. <u>Vaccine</u> - A weakened or inactive version of a pathogen that stimulates the body's production of antibodies which can destroy the pathogen

c. <u>Hormone</u> - A chemical released in the bloodstream that sends signals to specific cells, causing them to change their behavior in specific ways

2. a. <u>The lymphatic system fights disease.</u>

b. <u>The urinary system regulates water balance and chemical levels in the body.</u>

c. <u>The endocrine system controls various functions by releasing hormones.</u>

3. The lymphatic system is made up of <u>lymph vessels and lymph nodes</u>. The lymph gets cleaned in the <u>lymph nodes</u>.

4. <u>The contraction of certain muscles squeezes the lymph vessels, pumping lymph throughout the system.</u>

5. <u>The lacrimal glands produce tears. Tears clean the eye of contaminants and provide a chemical relief for sadness.</u>

6. The lymph nodes have <u>T-cells, B-cells, and macrophages</u> which all fight pathogens and toxic chemicals in different ways.

7. <u>B-cells</u> produce antibodies.

8. <u>Memory B-cells</u> give the lymphatic system a memory of past infections.

9. <u>It does little good</u>. The purpose of a vaccine is to give you the antibodies and the memory B-cells *before* the infection takes place.

10. <u>The renal artery brings blood into the kidney. This blood supplies oxygen and nutrients to the cells of the kidneys and picks up the cells' waste products. Then, the blood is filtered. The water and chemicals are dumped into tubes called nephrons. As the chemicals and water travel through the nephrons, cells absorb specific amounts of water and chemicals which get put back into the blood. Any water and chemicals left over get sent to the renal pelvis and out of the kidney. The cleaned blood leaves through the renal vein.</u>

11. <u>Excess water and chemicals are dumped into the renal pelvis and then travel through the ureter to the bladder. Eventually, they leave the body through the urethra.</u>

12. <u>Dialysis is the process by which a person is hooked up to an artificial kidney when their own kidneys are not functioning properly</u>.

13. <u>d</u>

14. <u>a</u>

15. <u>e</u>

16. <u>b</u>

17. <u>c</u>

18. The <u>pituitary gland</u> is often called the "master endocrine gland."

SOLUTIONS TO THE MODULE #16 STUDY GUIDE

1. a. <u>Autonomic nervous system</u> - The system of nerves which carries instructions from the CNS to the body's smooth muscles, cardiac muscle and glands

b. <u>Sensory nervous system</u> - The system of nerves which carries information from the body's receptors to the CNS

c. <u>Motor nervous system</u> - The system of nerves which carries instructions from the CNS to the skeletal muscles

2. <u>Neurons and neuroglia</u> are the two principal kinds of cells in the human nervous system.

3. *See figure 16.3:*
      a. <u>dendrite</u>   b. <u>nucleus</u>   c. <u>cell body</u>   d. <u>axon</u>

4. <u>Dendrites carry electrical signals to the cell body.</u>

5. <u>Axons carry electrical signals away from the cell body.</u>

6. <u>A synapse is a small gap between the axon of a neuron and the receiving end of another cell.</u>

7. <u>When the electrical signal reaches the end of the axon, neurotransmitters are released which travel across the synapse. Once they reach the receiving cell, they create a new electrical signal.</u>

8. <u>Neuroglia support the neurons by performing tasks which make it possible for the neurons to do their job.</u>

9. <u>The nerve is a part of the PNS.</u> The PNS contains all nerves that run off of the spinal cord. If it is in the leg, it is off the spinal cord.

10. <u>The CNS is made up of the brain and the spinal cord.</u>

11. <u>The skull and the cerebrospinal fluid protect the brain.</u>

12. <u>The vertebral column protects the spinal cord.</u> You could also say "the backbone."

13. <u>Gray matter is made up mostly of neuron cell bodies.</u>

14. <u>White matter is made up mostly of the axons of neurons.</u>

15. <u>The corpus callosum allows the two sides of the brain to communicate with one another.</u>

16. <u>The cerebellum is mostly in charge of voluntary muscle movements</u>, especially fine movements.

17. The cerebrum is in charge of most higher level thinking skills.

18. The two sides of the brain do not do exactly the same things. The left side of the cerebrum, for example, tends to be responsible for speaking, logic, and math. The right side is more involved with spatial relationships, recognition, and music.

19. The left side of the brain controls the right side of the PNS.

20. The blood-brain barrier is a system that insulates the brain from the blood. It is important because many of the chemicals in our blood are toxic to brain cells. The blood-brain barrier selectively transports "good" chemicals into the brain and leaves the "bad" chemicals in the capillaries, insulated from the brain.

21. The sympathetic division increases the rate and strength of the heartbeat and raises the blood pressure. It also stimulates the liver to release more glucose in the blood, producing quick energy for the "fight or flight" response that we experience when we are frightened or angry.

22. The parasympathetic system slows the heart rate and lowers the blood pressure. In addition, it takes care of certain "housekeeping" activities such as causing the stomach to churn while it is digesting a meal.

23. Humans detect four basic tastes: salty, sour, sweet, and bitter.

24. The tongue is least sensitive to taste at the center, because there are few taste buds there.

25. When we smell, we are actually detecting chemicals which are in the air.

26. The pupil regulates how much light gets into the eye.

27. The lens focuses light on the retina.

28. The ciliary muscle deforms the lens, changing its focus.

29. The rods and cones detect light.

30. Where the optic nerve enters the eye, there are no rods and cones. Without the rods and cones, light cannot be detected on that part of the retina.

31. Experiment 16.6 demonstrated that the fingers are more touch sensitive, which means they have more touch-related nerves.

32. The ear drum converts vibrations in the air into vibrations of the ossicles, which moves the cochlea back and forth.

33. <u>The cochlea converts the back and forth motion of the ossicles into electrical signals which can be received by the brain and interpreted as sound</u>.

Tests

TEST FOR MODULE #1

1. Define the following terms:

a. Science
b. Papyrus
c. Spontaneous generation

<u>MATCHING</u>
Match the person on the left with the proper description on the right

2. James Clerk Maxwell   a. Destroyed the idea of the immutability of species

3. Thales   b. Demonstrated the First Law of Thermodynamics

4. Galileo   c. He is best known for his model of the atom. It was named after him, and it revealed many of the atom's mysteries.

5. Ptolemy

6. Einstein   d. Discovered the Law of Mass Conservation

e. One of the first scientists

7. Lavoisier

f. Ancient Greek scientist who believed in atoms

8. Darwin

g. He developed the first detailed atomic theory and became known

9. Democritus   as the founder of modern atomic theory.

10. Grosseteste   h. Determined how traits are passed on during reproduction

11. Copernicus   i. Considered the first modern scientist

12. Aristotle   j. Had two theories of relativity and was big in quantum mechanics

13. Newton   k. Founder of modern physics

14. Mendel   l. Developed the idea of spontaneous generation

15. Joule   m. Proposed heliocentric system

16. Dalton   n. Proposed geocentric system

17. Bohr   o. The single greatest scientist of all time

p. Collected much data in favor of the heliocentric system but was forced to recant belief in it

18. What lesson can we learn from the fact that scientific progress stalled during the Dark Ages?

19. What caused scientific progress to move forward again towards the end of the Dark Ages?

20. What lesson can we learn from the fact that the idea of spontaneous generation was believed for so long, despite the evidence against it?

TEST FOR MODULE #2

1. Define the following terms:

a Counter-example
b. Hypothesis
c. Theory
d. Scientific law

2. Put the following steps of the scientific method into their proper order:

a. Theory is now a law
b. Hypothesis is now a theory
c. Make observations
d. Perform experiments to confirm the hypothesis
e. Form a hypothesis
f. Perform many experiments over several years

3. What is wrong with the following statement?

All objects, regardless of their weight, fall at the same rate

Questions 4-7 refer to the following story:

A slightly eccentric student is standing by a pool counting his pennies. He drops a penny and notices that it sinks to the bottom of the pool. He decides that all solid objects sink in water. Thus, the student starts dropping objects into the pool. He drops all of his coins, some rocks, books, a chair, and his shoes in the pool. All of them sink. He then proudly states that he has come up with a theory: All solid objects sink in water. Another student drops a cork into the pool and it floats. The eccentric student is crestfallen.

4. Did the eccentric student follow the scientific method?

5. If you answered "yes" to question 4, list the observation, hypothesis, and experiment designed to confirm the hypothesis. If you answered "no," explain why.

6. What did the other student provide to destroy the eccentric student's theory?

7. How is this story similar to the story about the theory that there are canals on Mars?

8. When scientists discovered high-temperature superconductors, it was quite surprising. Why?

9. What are the three limitations of science?

10. There is a lot of interest in how life originated on this planet. Can such a subject be studied by science?

TEST FOR MODULE #3

1. Define the following terms:

a. Experimental variable
b. Control (of an experiment)
c. Blind studies
d. Double-blind studies

2. Why is it important to analyze an experiment for experimental variables?

Questions 3 - 6 refer to the following story:

A consumer lab decides to test the claims of certain automatic dishwasher additives. Three brands claim to reduce the amount of water spots on dishes when you add them to your dishwasher in addition to your normal dishwashing liquid. The lab does a load of dirty glasses with a standard brand of dishwashing detergent with no additives. The lab then does three more loads, each time adding a different brand of additive to the same standard detergent. Each load of glasses is different, but they are all done in the same dishwasher, one after the other. Each load is inspected, and the cleanest load is determined.

3. Which of the following are experimental variables?

a. The dishwasher used
b. The additive used
c. The glasses used
d. The detergent used
e. The cleanliness of the dishwasher before each load is washed

4. From which of the experimental variables you identified above will the lab learn something?

5. Which of the experimental variables you identified above should be reduced or eliminated?

6. Is the data collected from this experiment subjective or objective?

7. Two things are floating on the surface of a sink full of water: a cork and a metal paper clip. What happens to the two items when dishwashing soap is put in the water?

8. A certain nutritional bar is supposed to give runners an extra "boost," allowing them to run without getting as tired as they normally do. To test the effectiveness of this nutritional bar, a scientist gets a group of volunteers together. She decides to give half of the volunteers the nutritional bar and the other half a regular granola bar. She then makes them run 3 miles and tell her whether they feel more or less tired than they usually do after a 3-mile run. Should this study be done blind, double-blind, or neither?

Questions 9 - 11 refer to the following story:

To investigate how well lead protects against radioactivity, a scientist puts a radiation detector near a radioactive metal. He then puts varying amounts of lead in between the radiation detector and the radiation source and measures how much radiation reaches the detector. His results are as follows:

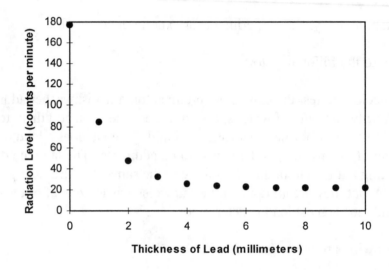

9. What is the radiation level that the detector sees when no lead is used to block the radiation?

10. If the scientist wants to reduce the radiation level to about 50 counts per minute, how many millimeters of lead should he use?

11. You are designing shielding for a lab. Your budget is tight, however. You want as much protection as possible, but you don't want to buy any more lead than is necessary. How many millimeters of lead should you use in your shielding?

TEST FOR MODULE #4

1. Define the following terms:

a. Simple machine
b. Force
c. Mechanical advantage
d. Diameter
e. Circumference

2. Which of the following experiments would be considered applied science experiments?

a. An experiment to find a better gasoline mixture for automobiles
b. An experiment to determine how to reduce pollutants produced by automobiles
c. An experiment to determine the level of pollutants in the air
d. An experiment to find the source of pollutants in the air

3. Which of the following are examples of technology?

a. A detailed description of the eating habits of bears
b. A list of substances which are poisonous to bears
c. A gun for hunting bears
d. A spray that, when sprayed around a campsite, keeps bears away

4. If you cannot generate enough force to lift a rock with a first-class lever the way you have it set up, should you change the setup so that the fulcrum is closer to or farther from the rock?

5. Scissors are an example of two levers put together. To which class do the levers belong?

6. If a third-class lever has a mechanical advantage of 5, what does that mean?

7. What is the mechanical advantage of a wheel and axle system whose wheel has a diameter of 20 inches, and whose axle has a diameter of 2 inches?

8. If you turn the wheel of a wheel and axle, what does the mechanical advantage do for you? What is the drawback that accompanies the mechanical advantage?

9. A block and tackle uses two pulleys. What is the mechanical advantage?

10. In the diagram below, is the simple machine an inclined plane or a wedge?

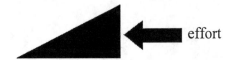 effort

11. If the slope of the machine in problem #10 is 10 inches and the height is 2 inches, what is the mechanical advantage?

12. What is the mechanical advantage of a screw that has a pitch of 0.05 inches and a head diameter of 0.1 inches?

13. If you turn the screw in problem #12 with a screwdriver whose diameter is 1 inch, what is the mechanical advantage?

TEST FOR MODULE #5

1. Define the following terms:

a. Life science
b. Archaeology
c. Geology
d. Known age
e. Dendrochronology
f. The Principle of Superposition

2. When testing for the historical accuracy of a document, which test evaluates whether or not the document we have today is the same document as the original?

3. When testing for the historical accuracy of a document, which test evaluates whether or not the document contradicts itself?

4. When testing for the historical accuracy of a document, which test evaluates whether or not the document contradicts other known historical or archaeological facts?

5. Although the genealogies of Jesus in Luke 3 and Matthew 1 seem to contradict each other, they really do not. Why? (You can use a Bible if you like.)

6. Of the three tests for a document of history, which two does the Bible pass better than any other document of its time?

7. Why must we apply Aristotle's dictum when using the internal test?

8. An archeologist finds a preserved log that was used to build an old shack. The rings of the log are clearly visible. In order to determine the age of the shack by dendrochronology, what must the archaeologist find in those rings?

9. Which type of age is more certain: a known age or an absolute age?

10. What kind of ages does dendrochronology provide, known ages or absolute ages?

11. If an artifact has neither a known age or an absolute age, what principle might be used to determine its age relative to something else?

12. The principle mentioned in problem #11 makes a very important assumption. What is that assumption?

TEST FOR MODULE #6

1. Define the following terms:

a. Catastrophism
b. Uniformitarianism
c. Humus
d. Minerals
e. Weathering
f. Erosion
g. Unconformity

2. Which hypothesis (uniformitarianism or catastrophism) can allow for a "young" earth?

3. A rock is formed and then later on is changed as a result of extreme pressure. What kind of rock is it now?

4. A layer of rock is laid down by water. What kind of rock is it?

5. A layer of rock is formed from lava that erupts from a volcano. What kind of rock is it?

6. An igneous rock is laced with iron. When exposed to water, the iron rusts and the rock crumbles. Is this chemical or physical weathering?

7. A river flows through 2 regions of the country. The first region has few plants, while the second region is covered with thick grass, flowers, and many trees. In which region do you expect the most erosion to occur?

8. A cavern has no groundwater seeping through its ceiling. Will you see stalactites in it? Will you see stalagmites in it?

(TEST CONTINUES ON THE NEXT PAGE)

Questions 9 and 10 refer to the diagram below:

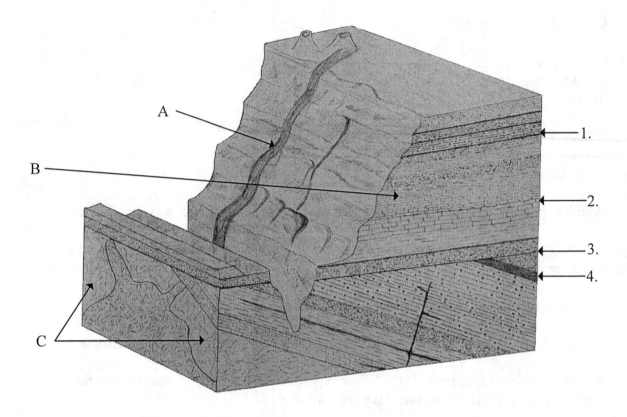

9. Identify the *type* of rock pointed out by each *letter* in the figure.

10. Which *number* in the figure points to the angular unconformity?

TEST FOR MODULE #7

1. Define the following terms:

a. Petrifaction
b. Resin
c. Extinct

Questions 2-6 refer to the following fossil types:

Molds
Casts
Petrified remains
Carbonate residues
Impressions
Creatures trapped in amber
Creatures trapped in ice

2. Which types of fossils have at least some of the original creature's remains preserved?

3. Which two kinds of fossils preserve only the shape and outer details of the creature?

4. Which two fossil types typically give the most detail about the original creature?

5. Which two fossil types require a lot of pressure in order to form?

6. Which fossil type requires mineral-rich water in order to form?

7. What are the four general features of the fossil record?

8. A geologist believes that every layer in a series of stratified rock represents a different time period in earth's past. Is this geologist a uniformitarian or a catastrophist?

9. A geologist believes that most fossils were all laid down over a short time period in earth's past. Is this geologist a uniformitarian or a catastrophist?

10. Is it possible to believe that the earth is only a few thousand years old if you are a uniformitarian?

TEST FOR MODULE #8

1. Define the following terms:

a. Index fossils
b. Geological column
c. The Theory of Evolution

Problems 2-5 refer to the figure below, which illustrates two different geological formations which a geologist is studying. The letters are just labels for the strata, and the symbols represent index fossils in each layer.

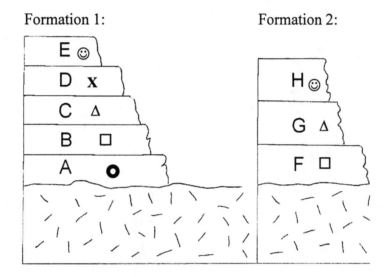

2. Using uniformitarian assumptions, identify each layer in formation 1 that represents the same time period as each layer in formation 2.

3. Considering both formations, which layer (A-H) would uniformitarians say is the very oldest layer of rock?

4. According to uniformitarian assumptions, which time periods represented by the layers in formation 1 are not in formation 2?

5. According to uniformitarians, which creatures came first: the ones whose fossils are represented by squares or the ones whose fossils are represented by smiley faces?

6. Why is the data from Mt. St. Helens considered evidence for catastrophism?

7. Is the geological column a real thing?

8. Does the fossil record support the idea of evolution or the idea that God created each plant and animal individually?

9. What is a paraconformity? Which viewpoint (uniformitarian or catastrophist) requires the existence of paraconformities?

10. Name two problems with uniformitarianism.

11. Name two problems with catastrophism.

TEST FOR MODULE #9

1. Define the following terms:

a. Atom
b. Molecule
c. Photosynthesis
d. Metabolism
e. Receptors
f. Cell

2. Here are two criteria for life:

- All life forms contain DNA.
- All life forms can sense changes in their surroundings and respond to those changes.

Which two are missing?

3. In the following drawing:

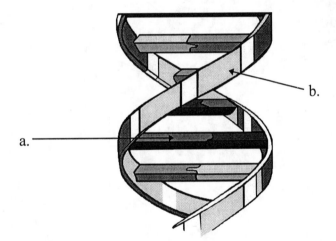

Which arrow is pointing to the backbone and which is pointing to a nucleotide?

4. How is information stored in DNA?

5. One half of a portion of DNA has the following sequence:

thymine, cytosine, guanine, adenine, guanine, thymine

What is the sequence of nucleotides on the other half of this portion?

6. What two things (use their proper, chemical names) are produced by photosynthesis?

7. Fill in the blank: Metabolism produces energy, carbon dioxide and _____.

8. If a certain type of organism produces lots of offspring, what can you conclude about the level of danger that the organism experiences throughout the course of its life?

9. Even if a person never has children, that person still reproduces. How?

10. In the following picture:

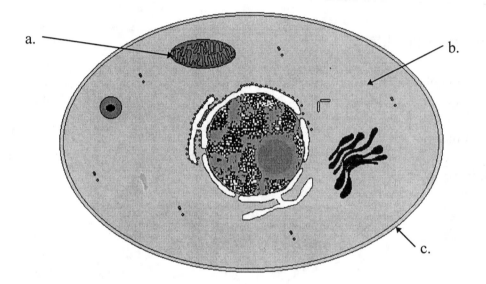

Indicate what the arrows are pointing out.

TEST FOR MODULE #10

1. Define the following terms:

a. Prokaryotic cell
b. Eukaryotic cell
c. Pathogen
d. Decomposers
e. Vegetative reproduction

2. Which kingdom is missing from the following list?

Monera, Plantae, Fungi, Animalia

3. An organism is made of many eukaryotic cells and eats other, living organisms. To which kingdom does it belong?

4. An organism is composed of one prokaryotic cell and eats other, living organisms. To which kingdom does it belong?

5. An organism is made of many eukaryotic cells and makes its own food. It has roots, stems, and leaves. To which kingdom does it belong?

6. An organism is made up of one eukaryotic cell and eats only dead organisms. To which kingdom does it belong?

7. Three strips of bacon are left lying on a table for 2 hours. The first is left uncovered. The second is salted. The third is salted and covered with plastic wrap.

    a. Which strip of bacon will be most contaminated by bacteria?

    b. Which will be least contaminated by bacteria?

8. What is the main difference between algae and plants? Which kingdom contains algae, and which kingdom contains plants?

9. Are decomposers an essential part of Creation?

10. A man walks out into his back yard and sees several mushrooms growing. He doesn't want mushrooms in his yard, so he picks all of the mushrooms, tearing them off at ground level. Afterwards, he sees no more mushrooms in his yard. The next day, he is horrified to see more. He picks all of them just as he did the day before. The next day, he sees more mushrooms again. Why do the mushrooms keep re-appearing?

11. When a plant wilts, does that mean it is dead?

12. What two parts of a plant cell work together to make turgor pressure?

TEST FOR MODULE #11

1. Define the following terms:

a. Axial skeleton
b. Appendicular skeleton
c. Exoskeleton
d. Symbiosis

For questions 2-11, Match the word on the left with the phrase that best describes it on the right.

2. smooth muscles	a. Makes bones flexible
3. skeletal muscles	b. Process by which cells are hardened and die in order to make hair, nails, and the outer layer of skin
4. keratinization	c. Make bones hard
5. bone marrow	d. Cushions the bones in a joint so that they do not rub together painfully
6. collagen	e. Involuntary muscles
7. minerals	f. Holds bones together in a joint
8. arthropods	g. Connects skeletal muscles to the skeleton
9. ligament	h. Animals with exoskeletons
10. cartilage	i. Voluntary muscles
11. tendon	j. Makes blood cells

12. Consider two joints. The first has a large range of motion while the second's range of motion is limited. Which, most likely, is the more stable joint?

13. If a person's biceps could not relax but was constantly contracted, would his forearm be flexed or extended?

14. A student experiments with 2 plants. The first one he waters but keeps in a dark closet. The second one gets watered and is also in a dark closet. The closet in which this plant is kept, however, has a window through which light can enter the closet. In three days, the first plant is dead and the second one has grown so that its leaves face the window. Which plant demonstrates phototropism?

15. A person's sweat glands are malfunctioning so that the person never sweats. Why is this person more likely to get sick than a person whose sweat glands are working?

16. Classify the following animals as mammal, reptile, amphibian, or bird.

      a. An ostrich – a creature that has feathers but cannot fly
      b. A brown bear with a thick, fur coat

TEST FOR MODULE #12

1. Define the following terms:

a. Producers
b. Consumers
c. Herbivore
d. Carnivore
e. Omnivore
f. Basal metabolic rate

2. Label each of the following as a consumer, producer, or decomposer:

a. yeast   b. ant   c. fly   d. corn   e. rosebush

3. Fill in the blanks:

The process of combustion requires _____ and makes _____, _____, and _____.

MATCHING: In numbers 4-10, match the word with the best description on the right:

4. carbohydrates   a. Usually the second macronutrient that is burned by the body

5. monosaccharides   b. The longest of the three basic kinds of carbohydrates

6. mitochondrion   c. An organism that does not have a constant internal temperature

7. fats   d. Usually the last macronutrient that is burned by the body

8. proteins   e. Usually the first macronutrient that is burned by the body

9. polysaccharides   f. The powerhouse of the cell

10. ectothermic   g. What carbohydrates must be broken down into before they are burned.

11. Why do endothermic organisms have a higher BMR than ectothermic organisms?

12. When a person doesn't eat enough protein or doesn't eat the right kinds of protein, what can the cells no longer do properly?

13. If you use lard or shortening (fats which are solid at room temperature) while you are cooking, are you using saturated fats or unsaturated fats?

TEST FOR MODULE #13

1. Define the following terms:

          a. Digestion   b. Vitamin

2. Identify the organs pointed out by the arrows labeled with arrows. Use the names at the left of the figure:

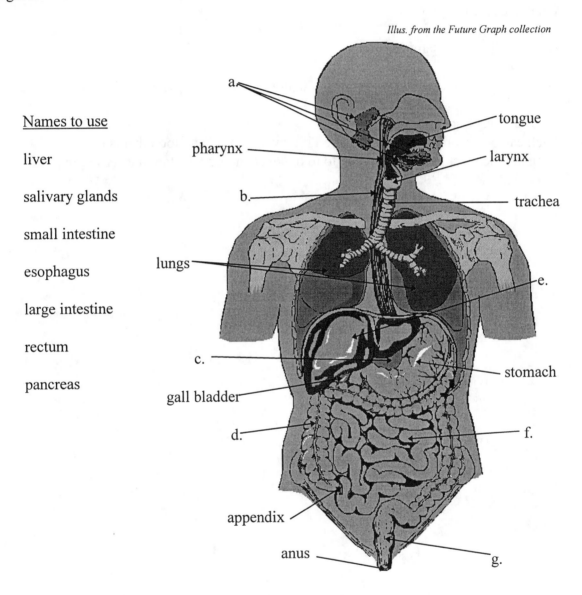

Illus. from the Future Graph collection

Names to use

liver

salivary glands

small intestine

esophagus

large intestine

rectum

pancreas

3. In which organ does most of the absorption of nutrients occur?

4. In which organ is the bolus turned into chyme?

5. In which organ is the undigested food turned into feces?

6. Which organ produces the sodium bicarbonate that neutralizes the stomach acid in the chyme?

7. Which organ produces bile?

8. Which organ produces saliva?

9. Which organ moves the food around in the mouth to form the bolus?

10. What does the epiglottis do?

11. For the following vitamins:

vitamin D, vitamin C, vitamin B_6, vitamin E, vitamin K

a. Which are water-soluble?
b. Which can be absorbed by the body even if they are not in any food that is eaten?
c. Which are the most likely to build up to toxic levels if you take too many vitamin pills?

TEST FOR MODULE #14

1. Define the following terms:

a. Veins
b. Arteries
c. Capillaries

2. Starting with the vena cava dumping a sample of deoxygenated blood into the heart, name each chamber of the heart in the order in which that sample will pass through it.

3. From which chamber of the heart does deoxygenated blood leave on its way to the lungs?

4. If a sample of blood is oxygenated, did it most likely come from an artery or a vein? Can you be 100% certain?

5. What makes up the majority of the blood: red blood cells, white blood cells, platelets, or plasma?

6. What is blood coagulation? What cells in the blood aid that process?

7. What blood cells contain hemoglobin?

8. What blood cells fight disease-causing organisms in the blood?

9. You are given a sample from a person's lungs. Looking at it under the microscope, you see many little round sacs that are covered in capillaries. What structures are you looking at?

10. When you *exhale* air, does the air pass through the trachea or the pharynx first?

11. You are listening to a man and a woman singing. The man is singing loudly with a very low pitch. The woman is singing much more softly but at a high pitch. Which singer is passing more air over his or her vocal cords?

12. Suppose you are looking through a microscope at a sample of living tissue. You watch the blood vessels and notice that the blood is going from oxygenated to deoxygenated as it passes through the vessels. What kind of blood vessels are you looking at: arteries, veins, or capillaries?

13. What are xylem? What is their purpose?

TEST FOR MODULE #15

1. Define the following terms:

a. Gland
b. Vaccine
c. Hormone

2. Which of the human systems is chiefly responsible for regulating water and chemical levels in the fluids of the body?

3. Which of the human systems is chiefly responsible for fighting disease?

4. In what structures can you find most of the infection-fighting cells of the lymphatic system?

5. What pumps lymph through lymph vessels?

6. What cells in the lymphatic system produce antibodies?

7. What cells in the lymphatic system give vaccines the ability to make people mostly immune from certain diseases?

8. A doctor has two medicines that fight the same disease. The first medicine is given if the patient has the disease. The second is given to patients who have never had the disease. Which of the two is a vaccine?

9. Once blood enters the kidney through the renal artery, it delivers oxygen and nutrients to the kidney's cells and picks up their waste products. The following steps then happen. Put them into the proper order:

    a. Chemicals and water are absorbed into the blood.
    b. The blood is filtered.
    c. Water and chemicals go into the renal pelvis, through the ureter, and into the bladder.
    d. Nutrients, water, and chemicals travel through a nephron.

Match the following glands with their function.

10. hypothalamus   a. Produces cortisol which causes the liver to release glucose into the blood
11. thyroid   b. Controls the pituitary gland
12. pituitary   c. Produces insulin which enables glucose to enter the cells
13. adrenal   d. Produces hormones that control many of the endocrine glands
14. pancreas   e. Produces hormones that regulate the basal metabolic rate

15. If many of the endocrine glands in the body begin to malfunction, just one gland might be responsible. Which gland might that be?

TEST FOR MODULE #16

1. Define the following terms:
   a. Autonomic nervous system
   b. Sensory nervous system
   c. Motor nervous system

Match the following structures with their appropriate description

2. neurons	a. carry signals towards a neuron's cell body
3. neuroglia	b. composed of all nerves running off of the spinal cord
4. dendrites	c. composed of cell bodies, dendrites, and axons
5. axons	d. allows the different hemispheres of the brain to communicate
6. synapse	e. sensitive to salty, bitter, sweet, and sour
7. neurotransmitters	f. converts the rocking motion of the ossicles into electrical signals that the brain interprets as sound
8. central nervous system	g. support the neurons by performing various tasks so that the neurons can do their job
9. peripheral nervous system	h. controls most of the high level thinking skills such as reasoning
10. gray matter	i. the part of the autonomic nervous system which speeds up the heart rate
11. corpus callosum	j. deforms the lens in the eye to adjust focus
12. cerebellum	k. chemicals which travel across the synapse, transmitting a signal from the end of one axon to a receiving cell
13. cerebrum	l. a gap between the axon of a neuron and the receiving cell
14. sympathetic division	m. the part of the autonomic nervous system which slows the heart rate
15. parasympathetic division	n. controls the movement of voluntary muscles
16. taste buds	o. composed of the brain and the spinal cord
17. cochlea	p. composed mostly of neuron cell bodies
18. ciliary muscle	q. carry signals away from a neuron's cell body

Solutions To The

Tests

SOLUTIONS TO THE TEST FOR MODULE #1

1. a. <u>Science</u> - A branch of study dedicated to the accumulation and classification of observable facts in order to formulate general laws about the natural world

b. <u>Papyrus</u> - A primitive form of paper, made from a long-leafed plant of the same name

c. <u>Spontaneous generation</u> - The idea that living organisms can be spontaneously formed from non-living substances

2. <u>k</u>

3. <u>e</u>

4. <u>p</u>

5. <u>n</u>

6. <u>j</u>

7. <u>d</u>

8. <u>a</u>

9. <u>f</u>

10. <u>i</u>

11. <u>m</u>

12. <u>l</u>

13. <u>o</u>

14. <u>h</u>

15. <u>b</u>

16. <u>g</u>

17. <u>c</u>

18. <u>The progress of science depends on government and culture</u>.

19. <u>A Christian worldview</u> caused science to progress at the end of the Dark Ages.

20. <u>We should believe scientific ideas because of *evidence*, not because of the people who believe in them</u>.

SOLUTIONS TO THE TEST FOR MODULE #2

1. a. <u>Counter-example</u> - An example that contradicts a scientific conclusion

b. <u>Hypothesis</u> - An educated guess that attempts to explain an observation or answer a question

c. <u>Theory</u> - A hypothesis that has been tested with a significant amount of data

d. <u>Scientific law</u> - A theory that has been tested by and is consistent with generations of data

2. c. <u>Make observations</u>
   e. <u>Form a hypothesis</u>
   d. <u>Perform experiments to confirm the hypothesis</u>
   b. <u>Hypothesis is now a theory</u>
   f. <u>Perform many experiments over several years</u>
   a. <u>Theory is now a law</u>

3. <u>The statement does not mention air.</u> The proper form of the statement would be:

   In the absence of air, all objects, regardless of their weight, fall at the same rate.

4. <u>Yes, he did.</u> He made an observation, formed a hypothesis, and performed several experiments to test his hypothesis. It doesn't matter that his conclusion was wrong. He followed the scientific method.

5. <u>He observed that a penny sank in water. This led him to the hypothesis that all solid objects sink in water. He performed experiments in which he threw many objects into the water, all of which sank. Thus, the hypothesis was confirmed.</u>

6. The other student provided a <u>counter-example</u> to show that the theory was wrong.

7. This story is similar because <u>the eccentric student's theory as well as the theory that there were canals on Mars were both produced by the scientific method but were both wrong.</u>

8. Scientists were quite surprised because <u>a scientific law said that they should not exist.</u>

9. a. <u>It cannot prove anything.</u>
   b. <u>It is not 100% reliable</u>
   c. <u>It must conform to the scientific method</u>

10. <u>Yes</u>, science can be used to study any question, as long as the scientific method is used.

INSTRUCTIONS TO THE PARENT/HELPER IN EXPERIMENT 2.3

You need to make the flashlight stop working in some way. You will find several ideas below, as well as tips you can give the student if the student is stuck. Please give the tips to the student *only* if he or she is stuck. Once you have stopped the flashlight from working, give it to the student so that he or she can start the experiment.

1. (This one is relatively easy.) Put a piece of paper between the two batteries, so that the batteries no longer make electrical contact with one another. The student will probably figure this one out as soon as he or she takes the flashlight apart. The experiment that the student designs to check the hypothesis should involve removing the paper to see that the flashlight is working, and then putting it back in to see that it stops working again. This will ensure that it was the paper causing the problem.

2. (This one is a little harder.) Put a *single* dead battery in the flashlight. Since your flashlight probably uses at least 2 batteries, that means there will be at least one good battery in the flashlight. It will probably be easy for the student to form the hypothesis, but it will be much harder to test. The student's experiment to test the hypothesis would have to involve replacing the batteries one at a time to determine *which* was the dead battery. Simply replacing all of the batteries and getting the flashlight working is NOT ENOUGH. The student must determine WHICH battery was dead.

3. (This one is a little harder than the first two.) If you have a flashlight with a bad bulb, use that flashlight. If you want to break the bulb (you can do this by dropping the flashlight a few times) in the flashlight you are using, that would work, too. In order for the student to be successful, however, you must have access to a good bulb. The experiment that the student would do should involve putting in a new bulb to fix the flashlight and then putting the old bulb back in to see that the flashlight stops working again.

4. (This is the hardest one.) If you unscrew the top of the flashlight and look at where the top touches the battery, you will see a little metal clip. You can use masking tape to cover that clip and then use a marker to color it black. This will camouflage what you have done. When the student discovers the problem, the experiment to confirm that it is the problem should involve removing it and seeing that the flashlight works followed by putting it back on to see that the flashlight no longer works.

Of course, if you can think of other ways to sabotage the flashlight, please use them. Remember, however, that **it is not enough for the student to just get the flashlight working**. It is commonplace to take something that was broken apart, put it back together, and find that it is suddenly working again. Thus, if the student just fixes the flashlight, he or she will not know for sure that the hypothesis was correct. Something else might have happened to accidentally fix the flashlight. That's why the student must perform an experiment which *demonstrates* that the student correctly identified the problem.

SOLUTIONS TO THE TEST FOR MODULE #3

1. a. <u>Experimental variable</u> - An aspect of an experiment which changes during the course of the experiment

b. <u>Control (of an experiment)</u> - The variable or part of the experiment to which all others can be compared

c. <u>Blind studies</u> - Experiments in which the participants do not know whether or not they are a part of the control group

d. <u>Double-blind studies</u> - Experiments in which neither the participants nor the people analyzing the results know who is in the control group

2. <u>Experimental variables can affect the results of your experiment</u>. The experimental variable from which you can learn something should be kept, and the rest should be reduced or eliminated so that they do not throw off the results of the experiment.

3. <u>Items (b), (c), and (e) are experimental variables</u>. They used the same dishwasher and detergent throughout, so neither of those are variable. The problem specifically states that they used different glasses and different additives. Since they used the same dishwasher again and again, it could very well be dirty from a previous load.

4. <u>You will learn something from (b)</u>.

5. <u>Items (c) and (e) should be reduced or eliminated</u>.

6. <u>The data is subjective</u>. It depends on the opinions of those inspecting the glasses, since there is no number that you can put on the level of cleanliness of dishes.

7. <u>When soap is put in the water, the paper clip will sink, but the cork will not</u>. The paper clip floats because of surface tension. You know that because, since it is made of metal, it is more dense than water. The only way it could float, then, is by surface tension. As you learned in Experiment 3.4, soap reduces surface tension, and it would cause the paper clip to sink. The cork floats not because of surface tension, but because it is less dense than water. Thus, it will keep floating regardless of surface tension.

8. <u>This should be a double-blind study</u>. If the participants know that they are getting a supposed energy boost, it might make them feel better mentally. This could make them feel less tired, regardless of whether the stuff in the bar did anything for them. The investigator should be blind as well, because listening to what people say and interpreting it is a subjective thing.

9. <u>The radiation level without any lead is 180 counts per minute</u>. No lead would be the case when the dot is at 0 on the x-axis. That dot is on the far left, and it is at a y-axis value of 180.

10. 50 counts per minute is halfway between 40 and 60 on the y-axis. The dot that has that vertical position is essentially at <u>2 millimeters</u> on the x-axis. You could have said anything between 1 and a half and 2 and a half.

11. After a thickness of 4 millimeters, no matter how far to the right you go, the dots don't go down much further at all. Thus, you don't gain much more protection after <u>4 millimeters</u>, so you might as well not waste any more money. You could have said 5 or 6 as well.

SOLUTIONS TO THE TEST FOR MODULE #4

1. a. <u>Simple machine</u> - A device that reduces the amount of force needed to perform a task or changes the direction of a force

b. <u>Force</u> - A push or pull that changes the motion of an object

c. <u>Mechanical advantage</u> - The amount by which force or motion is magnified in a simple machine

d. <u>Diameter</u> - The length of a straight line that travels from one side of a circle to another, while passing through the center of a circle

e. <u>Circumference</u> - The distance around a circle, equal to 3.1416 times the circle's diameter

2. <u>Experiments (a) and (b) are applied science experiments</u>, because the goal is specifically to make something better.

3. <u>Items (c) and (d) are technology</u>, because they make life better in some way.

4. The mechanical advantage of a lever increases the closer the fulcrum is to the resistance. Thus, <u>move the fulcrum closer to the rock</u>.

5. You apply a force at the handle of the scissors. Thus, the effort is at one end. The resistance is the thing you are cutting. In between is the little rivet that does not move. That's the fulcrum. Thus, the fulcrum is between the resistance and the effort, so scissors are composed of two <u>first-class levers</u>.

6. It means that <u>the speed at which the resistance moves will be magnified by 5</u>.

7. A wheel and axle's mechanical advantage is given by:

Mechanical advantage = (circumference of the wheel) ÷ (circumference of the axle)

Mechanical advantage = 20 ÷ 2 = <u>10</u>

8. In a wheel and axle, turning the wheel magnifies the effort. Thus, <u>the force with which the wheel is turned is magnified by 10. The drawback is that you must turn the wheel more than the axle turns</u>. The student can also say that the axle turns more slowly than the wheel.

9. The mechanical advantage of a multiple-pulley system (which is what a block and tackle is) is equal to the number of pulleys. Thus, the mechanical advantage is <u>2</u>.

10. As diagrammed, this is a <u>wedge</u>. See Figure 4.7.

11. Whether the machine is an inclined plane or a wedge, the mechanical advantage is the same:

Mechanical advantage = (length of slope) ÷ (height)

Mechanical advantage = 10 ÷ 2 = 5

12. To determine the mechanical advantage of a screw, we must first know the circumference of what's being grabbed. Since no screwdriver is mentioned, we must assume it is the head of the screw:

Circumference = 3.1416 x (diameter)

Circumference = 3.1416 x 0.1 = 0.31416

Now we can use the mechanical advantage equation for a screw:

Mechanical advantage = (circumference) ÷ (pitch)

Mechanical advantage = 0.31416 ÷ 0.05 = 6.2832

13. The mechanical advantage is different now, because a screwdriver is being used. For that, we need to know the circumference of the screwdriver:

Circumference = 3.1416 x (diameter)

Circumference = 3.1416 x 1 = 3.1416

Now we can use the mechanical advantage equation for a screw:

Mechanical advantage = (circumference) ÷ (pitch)

Mechanical advantage = 3.1416 ÷ 0.05 = 62.832

SOLUTIONS TO THE TEST FOR MODULE #5

1.a. <u>Life science</u> – A term that encompasses all scientific pursuits related to living organisms

b. <u>Archaeology</u> – The study of past human life as revealed by preserved relics

c. <u>Geology</u> – The study of earth's history as revealed in the rocks that make up the earth

d. <u>Known age</u> - The age of an artifact as determined by a date printed on it or a reference to the artifact in a work of history

e. <u>Dendrochronology</u> - The process of counting tree rings to determine the age of a tree

f. <u>The Principle of Superposition</u> – When artifacts are found in rock or earth that is layered, the deeper layers hold the older artifacts.

2. <u>The bibliographic test</u>

3. <u>The internal test</u>

4. <u>The external test</u>

5. <u>One traces Mary's lineage while the other traces Joseph's.</u>

6. <u>The Bible passes the external and bibliographic tests better than any other document of its time.</u>

7. <u>Some passages may look like contradictions, but the apparent contradiction might be due to difficulties in translation.</u> Thus, a contradiction must be ironclad in order to make an ancient document fail the internal test.

8. <u>The archaeologist must find a master tree ring pattern in the log's rings.</u> That's the only way to determine how old the log is using dendrochronology.

9. <u>Known ages are more certain.</u>

10. All dating methods (except the Principle of Superposition) give <u>absolute ages</u>.

11. <u>The Principle of Superposition.</u>

12. <u>The Principle of Superposition assumes that in rock or soil that is layered, the layers were formed one at a time.</u> This is not necessarily true.

SOLUTIONS TO THE TEST FOR MODULE #6

1. a. Catastrophism - The view that most of earth's geological features are the result of large-scale catastrophes such as floods, volcanic eruptions, etc.

b. Uniformitarianism - The view that most of earth's geological features are the result of slow, gradual processes that have been at work for millions or even billions of years

c. Humus - The decayed remains of once-living creatures

d. Minerals - Inorganic crystalline substances found naturally in the earth

e. Weathering - The process by which rocks are broken down by the forces of nature

f. Erosion - The process by which sediments and rock fragments are carried away by wind or rain

g. Unconformity - A surface of erosion that separates one layer of rock from another

2. Catastrophism allows for a "young" earth, because in catastrophism, geological structures are formed rapidly.

3. The rock is metamorphic.

4. The rock is sedimentary.

5. The rock is igneous.

6. This is chemical weathering, because the iron changed to a new substance: rust.

7. The first region will experience more erosion, as plants tend to reduce the effect of erosion.

8. Without groundwater seeping in through the ceiling, you will see neither stalactites nor stalagmites. Remember, even stalagmites are formed from water that drips in from the ceiling.

9. A. igneous rock
    B. sedimentary rock
    C. metamorphic rock

10. The number 3 points to the angular unconformity.

SOLUTIONS TO THE TEST FOR MODULE #7

1. a. <u>Petrifaction</u> - The conversion of organic material into stone

b. <u>Resin</u> – A sticky, liquid substance that usually hardens when exposed to air

c. <u>Extinct</u> - A term applied to a species of plant or animal that was once living but now is not

2. <u>Creatures trapped in amber and creatures trapped in ice</u> both contain at least some of the original remains of the creature. The student could also say carbonate residues, although the film on a carbonate residue is usually not considered an original remain. Don't count petrified remains as wrong either, because some petrified remains do have preserved parts of the original remains.

3. <u>Casts and molds</u>

4. Typically, <u>creatures trapped in amber and creatures trapped in ice</u> provide the most detail, as they usually have many of the soft parts as well as the hard parts of the creature preserved.

5. <u>Impressions and carbonate residues</u>

6. <u>Petrified remains</u>

7. The four general features of the fossil record are:

1. <u>Fossils are usually found in sedimentary rock. Since most sedimentary rock is laid down by water, it follows that most fossils were laid down by water as well.</u>
2. <u>The vast majority of the fossil record is made up of clams and other hard-shelled creatures. Most of the remaining fossils are of water-dwelling creatures and insects. Only a tiny, tiny fraction of the fossils we find are of plants, reptiles, and mammals.</u>
3. <u>Many of the fossils we find are of plants and animals which are still alive today. Some of the fossils we find are of plants and animals which are now extinct.</u>
4. <u>The fossils found in one layer of stratified rock can be considerably different than the fossils found in another layer of the same stratified rock.</u>

8. <u>The geologist is a uniformitarian.</u> Although many catastrophists do believe that the lowest layers of rock represent the Creation Week and the time before the flood and that the upper layers represent mostly the flood, uniformitarians believe each and every layer was laid down during a different period of earth's history.

9. <u>The geologist is a catastrophist.</u> Catastrophists believe that most fossils were laid down during the flood.

10. <u>No</u>, you cannot believe the earth is only a few thousand years old if you are a uniformitarian. Uniformitarians need hundreds of millions of years for the sedimentary rocks we see today to be formed by the processes we see occurring today.

SOLUTIONS TO THE TEST FOR MODULE #8

1. a. <u>Index fossils</u> - A fossil considered to represent a certain period in earth's past

b. <u>Geological column</u> - A theoretical picture in which layers of rock from around the world are meshed together into a single, unbroken record of earth's past

c. <u>The Theory of Evolution</u> - A theory which states that all of life on this earth has a single, common ancestor that existed a long time ago

2. <u>Layer B and layer F represent the same time period. Layer C and layer G represent the same time period. Layer E and layer H represent the same time period.</u> You can tell this because they have the same index fossils.

3. Layer F is at the bottom of formation 2, so it is the oldest layer in that formation. Layer A is at the bottom of formation 1, so it is the oldest layer in that formation. However, the question asks which layer is the very oldest. Well, since layer H and layer B have the same index fossils, that means they represent the same time period. Since layer A is under layer B, that means it is older than layer B. This means A is older than F, because F represents the same time period as B. Thus, <u>A is the oldest layer.</u>

4. <u>The time periods represented by layers A and D are not in formation 2</u>, because the index fossils for those time periods are not in formation 2.

5. Since the squares appear lower in the formations than the smiley faces, uniformitarians would assume that <u>the creatures whose fossils are represented by squares came first.</u>

6. <u>The data from Mt. St. Helens indicate that the major geological formations we see today can be formed quickly as a result of catastrophes.</u>

7. <u>No.</u> The geological column is a theoretical construct.

8. <u>It supports the idea that God created each plant and animal individually.</u> It indicates that plants and animals are unique and there are no intermediate links between them. This is evidence that the plants and animals were created individually.

9. <u>A paraconformity is an unconformity that does not exist. But geologists nevertheless assume that it does. Uniformitarians require their existence</u> to deal with situations where index fossils from different time periods appear in a single layer of rock.

10. The text discusses at least five problems. The student need only list 2.

a. There are too many fossils in the fossil record.

b. Fossil graveyards with fossils from many different climates are hard to understand in the uniformitarian view.

c. Index fossils are called into question by the many creatures we once thought were extinct but now know are not.

d. Uniformitarians must assume the existence of paraconformities.

e. Uniformitarians must believe that evolution occurred, and there is no evidence for evolution. In fact, the fossil record provides evidence that each plant and animal was created by God.

11. The text discusses at three problems. The student need only list 2.

a. Catastrophists have offered no good explanation for the existence of unconformities between rock layers laid down by the flood.

b. Catastrophists cannot explain certain fossil structures that look like they were formed under "normal" living conditions which would not exist during the flood.

c. Catastrophists have not yet explained the enormous amount of limestone that exists all over the world in terms of Noah's flood.

SOLUTIONS TO THE TEST FOR MODULE #9

1. a. <u>Atom</u> - The smallest stable unit of matter in Creation

b. <u>Molecule</u> - Two or more atoms linked together to make a substance with unique properties

c. <u>Photosynthesis</u> - The process by which a plant uses the energy of sunlight and certain chemicals to produce its own food

d. <u>Metabolism</u> - The process by which a living organism takes energy from its surroundings and uses it to sustain itself, develop, and grow

e. <u>Receptors</u> - Special structures or chemicals that allow living organisms to sense the conditions of their surroundings

f. <u>Cell</u> - The smallest unit of life in Creation

2.
- <u>All life forms have a method by which they take energy from the surroundings and convert it into energy that helps them live.</u>

- <u>All life forms reproduce.</u>

3. <u>Arrow (a) is pointing to a nucleotide, and arrow (b) is pointing to the backbone.</u>

4. <u>Information is stored in DNA as a sequence of nucleotides.</u> Just like this entire book could be sent as a sequence of dots and dashes in Morse code, all of the information of life can be stored as a sequence of nucleotides.

5. Since only adenine and thymine link up, and since only cytosine and guanine link up, the other half of the DNA must be:

<u>adenine, guanine, cytosine, thymine, cytosine, adenine</u>

6. <u>Photosynthesis produces glucose and oxygen.</u>

7. <u>water</u>

8. <u>Organisms that produce lots of offspring tend to live very dangerous lives.</u> They need lots of offspring to "replace" those that die before having offspring.

9. <u>That person's cells are reproducing all of the time.</u> Thus, the person has reproduced countless times on the cellular level.

10. a. <u>an organelle</u>

b. <u>cytoplasm</u>

c. <u>membrane</u>

SOLUTIONS TO THE TEST FOR MODULE #10

1. a. <u>Prokaryotic cell</u> - A cell that has no organelles

b. <u>Eukaryotic cell</u> - A cell with organelles

c. <u>Pathogen</u> - An organism that causes disease

d. <u>Decomposers</u> - Organisms that break down the dead remains of other organisms

e. <u>Vegetative reproduction</u> - The process by which the stem of a plant can form new roots and develop into a mature plant

2. <u>Protista</u>

3. Since it is made of eukaryotic cells, it is not in kingdom Monera. Since it is neither one-celled nor does it make its own food, it is not in kingdom Protista nor is it in kingdom Plantae. It eats living organisms, so it is not in kingdom Fungi. It must be in kingdom <u>Animalia</u>.

4. What it eats is irrelevant. Since it is a single prokaryotic cell, it belongs in kingdom <u>Monera</u>.

5. Organisms that make their own food and have specialized structures like roots, stems, and leaves belong in kingdom <u>Plantae</u>.

6. Since it is a single eukaryotic cell, you might think it belongs in kingdom Protista. However, it is a decomposer because it eats only dead organisms. Thus, it is a single-celled member of kingdom <u>Fungi</u>.

7. Salt will reduce bacterial growth and reproduction, and covering will reduce the new bacteria introduced onto the bacon.

    a. The <u>first</u> will have the most bacteria.

    b. The <u>third</u> will have the least.

8. <u>Algae have no specialized structures, while plants do. Algae belong in kingdom Protista, while plants belong in kingdom Plantae.</u>

9. <u>Yes</u>, decomposers recycle dead matter so that it can be used by living organisms again.

10. <u>The man did not really get rid of the mushroom fungus. The mycelium is still underground, so mushrooms will continually be produced.</u>

11. <u>No</u>, a wilting plant is not dead. It is just losing turgor pressure. Add water, and the wilting will go away. Plants will wilt before they die, but just being wilted does not necessarily mean the plant has died. Many wilted plants live for a long time if they get water soon enough.

12. The <u>central vacuole</u> and the <u>cell wall</u> work together to make turgor pressure. The central vacuole expands by filling with water, pushing the contents of the cell against the cell wall. The cell wall pushes back, making the pressure.

SOLUTIONS TO THE TEST FOR MODULE #11

1. a. <u>Axial skeleton</u> – The portion of the skeleton that supports and protects the head, neck, and the trunk

b. <u>Appendicular skeleton</u> – The portion of the skeleton that attaches to the axial skeleton and has the limbs attached to it

c. <u>Exoskeleton</u> - A body covering, typically made of a tough, flexible substance called chitin, that provides support and protection

d. <u>Symbiosis</u> - Two or more different organisms living together so that each benefits from the other

2. <u>e</u>

3. <u>i</u>

4. <u>b</u>

5. <u>j</u>

6. <u>a</u>

7. <u>c</u>

8. <u>h</u>

9. <u>f</u>

10. <u>d</u>

11. <u>g</u>

12. The larger the range of motion, the less the stability of a joint. Thus, the <u>second joint is probably more stable</u>.

13. When the biceps contracts, the forearm rises and the elbow is bent. If the biceps cannot relax, the arm cannot extend. Thus, <u>his forearm will always be flexed</u>.

14. Phototropism is the tendency of plants to grow towards the light. That was exhibited by the <u>second plant</u>. The first plant simply demonstrates that plants need light to survive.

15. This person is more likely to get sick because sweat feeds bacteria and fungi which fight off pathogenic organisms. Without the sweat, these beneficial organisms would die, and the person will be more likely to be infected by pathogenic organisms.

16. a. If an animal has feathers, it is a bird. It is irrelevant whether or not the creature can fly. The presence of feathers means it's a bird.

b. If an animal has hair, it is a mammal.

SOLUTIONS TO THE TEST FOR MODULE #12

1. a. <u>Producers</u> - Organisms that produce their own food

b. <u>Consumers</u> - Organisms that eat living producers and/or other consumers for food

c. <u>Herbivore</u> - A consumer that eats producers exclusively

d. <u>Carnivore</u> - A consumer that eats only other consumers

e. <u>Omnivore</u> - A consumer that eats both plants and other consumers

f. <u>Basal metabolic rate</u> - The minimum amount of energy required by the body every day

2. a. Yeast are in kingdom Fungi, so they are <u>decomposers</u>.

b. An ant cannot make its own food, so it is a <u>consumer</u>.

c. A fly cannot make its own food, so it is a <u>consumer</u>.

d. Corn is a plant, so it is a <u>producer</u>.

e. A rosebush is a plant, so it is a <u>producer</u>.

3. Combustion requires <u>oxygen</u> and produces <u>carbon dioxide</u>, <u>water</u>, and <u>energy</u>.

4. <u>e</u>

5. <u>g</u>

6. <u>f</u>

7. <u>a</u>

8. <u>d</u>

9. <u>b</u>

10. <u>c</u>

11. <u>Endothermic organisms must expend a lot of energy keeping their body temperature high</u>. Since this is independent of activity, it is a part of the BMR.

12. Without the proper proteins, <u>a person's cells cannot manufacture the proteins they need to manufacture</u>. In order to make proteins, cells need amino acids. There are 8 amino acids our cells cannot make. Thus, we must eat them. If we don't, our cells run out and cannot make proteins that have those amino acids in them.

13. <u>Saturated fats</u> are solid at room temperature.

SOLUTIONS TO THE TEST FOR MODULE #13

1. a. <u>Digestion</u> - The process by which an organism breaks down its food into small units which can be absorbed

b. <u>Vitamin</u> - a chemical substance essential for the normal working of the human body

2. a. <u>salivary glands</u>

b. <u>esophagus</u>

c. <u>pancreas</u>

d. <u>large intestine</u>

e. <u>liver</u>

f. <u>small intestine</u>

g. <u>rectum</u>

3. <u>The small intestine</u> is where most nutrient absorption occurs.

4. <u>The stomach</u> turns the bolus into chyme.

5. <u>The large intestine</u> converts waste into feces.

6. <u>The pancreas</u> produces sodium bicarbonate.

7. <u>The liver</u> produces bile.

8. <u>The salivary glands</u> produce saliva.

9. <u>The tongue</u> moves the food in the mouth to form the bolus.

10. <u>The epiglottis covers the larynx when you swallow to make sure that food goes down the esophagus only.</u>

11. a. <u>Vitamins C and B$_6$</u>. Remember, A, D, E and K are the fat-soluble vitamins. If a vitamin is not fat-soluble, it is water-soluble.

b. <u>Vitamins D and K</u>. Vitamin D can be made by the body through sunlight hitting the skin, and vitamin K is made by bacteria in the large intestine.

c. <u>Vitamins D, E, and K</u>. The fat-soluble vitamins are the ones that build up easily in the body. These are the fat-soluble ones on the list.

SOLUTIONS TO THE TEST FOR MODULE #14

1. a. <u>Veins</u> - Blood vessels that carry blood back to the heart

b. <u>Arteries</u> - Blood vessels that carry blood away from the heart

c. <u>Capillaries</u> - Tiny, thin-walled blood vessels that allow the exchange of gases and nutrients between the blood and cells

2. Blood flows into the <u>right atrium</u> and then gets dumped into the <u>right ventricle</u>. It then goes to the lungs and returns oxygenated. The oxygenated blood is dumped into the <u>left atrium</u> and then gets dumped into the <u>left ventricle</u> before being pumped to the body.

3. Deoxygenated blood leaves the <u>right ventricle</u> on its way to the lungs.

4. Most likely, the blood came from an <u>artery</u>, because most arteries contain oxygenated blood. <u>You cannot be 100% certain</u>, however, because there are exceptions.

5. <u>Plasma</u> makes up the majority of the blood.

6. <u>Blood coagulation is the process by which blood clots to keep it from leaking out an injured artery. It is aided by the blood platelets</u>.

7. <u>Red blood cells</u> contain hemoglobin.

8. <u>White blood cells</u> fight disease.

9. You are looking at <u>alveoli</u>.

10. Exhaled air travels from the lungs, into the trachea, up the larynx, to the pharynx, and out the mouth or nasal cavity. Thus, it passes through the <u>trachea</u> first.

11. Volume is controlled by how much air passes over the vocal cords. Thus, <u>the man</u> is passing more air over his vocal cords than is the woman.

12. You are looking at <u>capillaries</u>. Those are the only vessels in which oxygen exchange occurs.

13. <u>Xylem are tubes in plants that transport water up the plant</u>.

SOLUTIONS TO THE TEST FOR MODULE #15

1. a. <u>Gland</u> - A group of cells that prepare and release a chemical for use by the body

b. <u>Vaccine</u> - A weakened or inactive version of a pathogen that stimulates the body's production of antibodies which can destroy the pathogen

c. <u>Hormone</u> - A chemical messenger released in the bloodstream that sends signals to distant cells, causing them to change their behavior in specific ways

2. <u>The urinary system</u>

3. <u>The lymphatic system</u>

4. The lymphocytes are found in the <u>lymph nodes</u>.

5. Lymph is pumped through lymph vessels by <u>the contraction of muscles in the body</u>.

6. <u>B-cells</u> make antibodies.

7. <u>Memory B-cells</u> give the lymphatic system a "memory" of infections. Vaccines use this memory feature to give the body immunity to diseases it hasn't actually fought yet.

8. <u>The second is the vaccine</u>. Vaccines are used to give a person immunity to a disease he or she has not had yet. Only in very rare cases can a vaccine be used as a treatment for a disease.

9. The order is <u>b, d, a, c</u>.

10. <u>b</u>

11. <u>e</u>

12. <u>d</u>

13. <u>a</u>

14. <u>c</u>

15. The <u>pituitary gland</u> might be malfunctioning. Since it controls most endocrine glands, a failure there would make most endocrine glands fail.

SOLUTIONS TO THE TEST FOR MODULE #16

1. a. <u>Autonomic nervous system</u> - The system of nerves which carries instructions from the CNS to the body's smooth muscles, cardiac muscle and glands

b. <u>Sensory nervous system</u> - The system of nerves which carries information from the body's receptors to the CNS

c. <u>Motor nervous system</u> - The system of nerves which carries instructions from the CNS to the skeletal muscles

2. <u>c</u>

3. <u>g</u>

4. <u>a</u>

5. <u>q</u>

6. <u>l</u>

7. <u>k</u>

8. <u>o</u>

9. <u>b</u>

10. <u>p</u>

11. <u>d</u>

12. <u>n</u>

13. <u>h</u>

14. <u>i</u>

15. <u>m</u>

16. <u>e</u>

17. <u>f</u>

18. <u>j</u>